Student Solutions Manual

Mathematics for Elementary School Teachers

SIXTH EDITION

Tom Bassarear
Keene State College

Meg Moss
Western Governors University

Prepared by

Laura Wheel

⁂ CENGAGE
Learning

Australia · Brazil · Mexico · Singapore · United Kingdom · United States

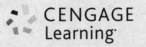

For product information and technology assistance, contact us at **Cengage Learning Customer & Sales Support, 1-800-354-9706**.

For permission to use material from this text or product, submit all requests online at **www.cengage.com/permissions** Further permissions questions can be emailed to **permissionrequest@cengage.com**.

ISBN: 978-1-305-10833-2

Cengage Learning
20 Channel Center Street
Boston, MA 02210
USA

Cengage Learning is a leading provider of customized learning solutions with office locations around the globe, including Singapore, the United Kingdom, Australia, Mexico, Brazil, and Japan. Locate your local office at: **www.cengage.com/global**.

Cengage Learning products are represented in Canada by Nelson Education, Ltd.

To learn more about Cengage Learning Solutions, visit **www.cengage.com**.

Purchase any of our products at your local college store or at our preferred online store **www.cengagebrain.com**.

Printed in the United States of America
Print Number: 01   Print Year: 2014

Contents

CHAPTER 1 Foundations for Learning Mathematics

SECTION 1.1: Getting Started and Problem Solving

1. Answers will vary.

3. Answers will vary.

5.

Student A:	Student B:
86	86
× 47	× 47
602	42
344	560
4042	240
	3200
	4042

7. You pay $4 \cdot 4 + \$0 \cdot 1 = \$16$ for 5 movies. Since $\$16 \div 5 = \$3.20$, you pay $3.20 per movie.

9. In one week, Sally makes 40 hours × $6.85 per hour = $274. Assuming she works 5 days a week, she pays $15 \cdot 5 = \$75$ per week for child care. After deducting child care expenses, she makes $274 − \$75 = \$199$. Therefore, she actually makes $199 \div 40 = \$4.975$, or $4.98 per hour.

11. a. If one penny has a diameter of 0.75 inches, then we have 288.7 × 0.75 = 216.525 billion inches; or 216.525 ÷ 12 = 18.04375 billion feet; or 18.04375 ÷ 5280 ≈ 0.00341740 billion miles = 3,417,400 miles.

    b. 30,000,000 ÷ 24 = 1,250,000 per hour; 1,250,000 ÷ 60 = 20,833 per minute; 20,833 ÷ 60 = 347 per second. 30,000,000 × 365 = 10,950,000,000 pennies per year.

13. a. Need to make assumptions for the distance driven and the average cost of gasoline.

    b. Answers will vary. Sample answer: If the distance is 2400 miles and the cost of gas averages $3.00 per gallon, then the cost of gas for each is as follows:

    Van: (2400 ÷ 18) × 3 = $400   Sedan: (2400 ÷ 32) × 3 = $225

    The difference is $400 − \$225 = \$175$. So, it will take $175 extra to drive the van.

    c. Using the assumptions in part b, if the price of gas rises 40¢, then it is now averaging $3.40 per gallon. Using the same procedure as in part b, it costs $53.33 more for the van and $30 more for the sedan.

15. In one minute, a hummingbird's wings beat 60 times per second × 60 seconds = 3600 times. In an hour, a hummingbird's wings beat 3600 times per minute × 60 minutes = 216,000 times.

17. Each side of the square 24 ÷ 4 = 6 toothpicks long. So, Karen used 6 × 6 = 36 toothpicks to make the hexagon.

1

SECTION 1.2: Process, Practice, and Content Standards: Getting Started

1. Answers will vary.

3. The tournament starts with a round of 4, which indicates the number of games played in this round. In the next round with 64 teams, 32 games are played. From this round, 32 teams move on to the next round in which 16 games are played, and so on until the final championship game.
    Therefore, $4 + 32 + 16 + 8 + 4 + 2 + 1 = 67$ games are played in the NCAA men's basketball tournament.

5. We don't know, because an answer of 3:15 assumes that the power was off only a matter of seconds. It is possible that the power went off at 1 A.M. and came back on at 3:15 A.M. Thus we can say only that the power came back on at 3:15 because electric clocks begin at 12 A.M. when the power comes back on. If it is 6:45 A.M., and the clock says 3:30, it means that the power came back on $3\frac{1}{2}$ hours ago; that is, at 3:15 A.M.

7. a.

$\dfrac{104-(8\times6)}{7} = \dfrac{104-48}{7} = \dfrac{56}{7} = 8$; She should hang each plate with an 8-inch space between each plate and between the end plates and the wall.

b. If one plate breaks, she would have 5 plates to hang.

$\dfrac{104-(8\times5)}{6} = \dfrac{104-40}{6} = \dfrac{64}{6} = 10\dfrac{4}{6} = 10\dfrac{2}{3}$; She should hang each plate with a $10\frac{2}{3}$-inch space between each plate and between each end plate.

c. $\dfrac{104-(12\times2)-(8\times6)}{5} = \dfrac{104-24-48}{5} = \dfrac{32}{5} = 6\dfrac{2}{5}$; She should hang each plate with a $6\frac{2}{5}$-inch space between them.

9. There are 8 tricycles and 24 bicycles.

11. The three possibilities are:

Bicycles	Tricycles
6	0
3	2
0	4

13.   There are 81 tribbles and 16 chalkas.

15.   **a.**   The possible scores are 4 through 22, 24, 25, 26, 28, and 32.

     b.   Two ways: $1 + 1 + 2 + 8$ and $2 + 2 + 4 + 4$

     c.   There must be an even number of ones for the result to be even. There are three ways to get an even number: no 1s, two 1s, or four 1s. Of course, the last possibility will not give us 12. If we use two 1s, then the remaining two throws must result in a total of 10. The only way to do this is $2 + 8$. Finally, using no 1s, suppose we have one 8-point throw, then we must score the remaining 4 points in three throws, which is impossible. Thus, an 8-point dart cannot be used, If we have a 4-point throw, then the remaining three throws must total 8 points. The only way to do this is $2 + 2 + 4$.

17.   **a.** $64 \div 3\frac{2}{3} = 17\frac{5}{11}$; so, 17 bottles.

     b. $64 - \left(17 \times 2\frac{2}{3}\right) = \frac{5}{3} = 1\frac{2}{3}$; so, $1\frac{2}{3}$ ounces (or about 1.7 ounces) of leftover perfumes per jug.

     c.   Any jug size that has a capacity that is a multiple of 11 ounces. For example, an 11-ounce jug.

19.   **a.**   Fill the 5-gallon pail and use it to fill the 3-gallon pail (leaving 2 gallons in the 5-gallon pail). Empty the 3-gallon pail and put the 2 gallons left in the 5-gallon pail into the 3-gallon pail. Fill the 5-gallon pail again and use it to finish filling the 3-gallon pail. This leaves 4 gallons in the 5-gallon pail.

     b.   Fill the 3-gallon jug from the 8-gallon jug and empty it into the 5-gallon jug. Fill the 3-gallon jug again (from the 8-gallon jug). This leaves 3 gallons in both the 3-gallon jug and 5-gallon jug and $8 - 6 = 2$ gallons in the 8-gallon jug.

21.   The ball bounces 4 times. After the first bounce, the ball rises $\frac{1}{2} \times 16 = 8$ feet. After the second bounce, the ball rises 4 feet; after the third bounce, the ball rises 2 feet; and after the fourth bounce the ball rises 1 foot.

23. $20 \times 19 = 380$ valentines. This is not the same as the handshake problem. If students A and B shake hands, then the handshake counts for both students A and B. If student A gives student B a valentine, it only counts for student A. Student B still has to give student A a valentine.

25. $a = 1474$; $b = 1628$; $c = 5115$

27.   8712

29.   Answers will vary.

31.   **a.**

9	19	5
7	11	15
17	3	13

     b.

4	18	8
14	10	6
12	2	16

     c.

1	26	0
8	9	10
18	−8	17

33.   **a.-b.**   Answers will vary.

35. a.

50	25	10	5	1
1	1	1	3	0
1	0	5	0	0
0	3	2	1	0

b.

50	25	10	5	1
1	0	3	3	0
0	3	0	4	0
0	2	4	1	0

c.

50	25	10	5	1
1	0	2	3	15
0	1	3	7	10
0	2	3	1	15
0	0	7	4	10
0	0	3	13	5

37. Answers will vary.

39. a. The sum of the first seven numbers is $1 + 1 + 2 + 3 + 5 + 8 + 13 = 33$, which is one less than the ninth term. The sum of the first eight terms is 54, which is one less than the tenth term. So, the sum of the first n terms is equal to one less than the $(n + 2)$nd term, or
$$A_1 + A_2 + A_3 + \cdots + A_n = A_{n+2} - 1.$$

b. $A_n \times A_{n+2} = (A_{n+1})^2 \pm 1$

41. Answers will vary. There are many possible answers.

43. 325

45. Yes. Use three 15¢ stamps and four 33¢ stamps.

47. Starting at 32, he circles every sixth number; that is, he adds 6 to get the next number. He forgot to circle 56 and 62.

49. 6 tables: 4 round and 2 square

51. If all dogs got 20 votes, there would be a 4-way tie. If one dog gets 21 votes, then it will win the contest. The other three dogs could get 20, 20, and 19 votes.

CHAPTER 1 REVIEW EXERCISES

1. 316 student tickets

3. 15 coins equaling 92 cents

50	25	10	5	1
1	1	0	1	12
1	0	0	7	7
0	1	1	11	2
0	1	5	2	7
0	2	1	5	7
0	0	5	8	2

5. Nine ways.

Quarter	Dimes	Nickels
1	2	1
1	1	3
1	0	5
0	5	0
0	4	2
0	3	4
0	2	6
0	1	8
0	0	10

7. You pay $1.30 \cdot 9 + \$0 \cdot 1 = \$11.70$ for 10 cups of coffee. Since $\$11.70 \div 10 = \$1.17$, you pay $1.17 per cup of coffee.

9. There were two 10¢ stamps and ten 5¢ stamps.

11. The pattern repeats after every 12th letter; so, the 240th letter in the pattern is Y. Since 243 is 3 more than 240, the 243rd letter in the pattern is P.

13. a. The sum of the numbers along the length of the stick equals the number at the end of the stick.

    b. Since the largest number on the chart is 924, working backwards: 1, 6, 21, 56, 126, 252, 462.

    c. The handle would be the diagonal row of 13 ones along the left edge of the chart, and at the end of the stick would be 12.

    d. Answers will vary.

15. a.   Fill the 9-gallon pail and use it to fill the 4-gallon pail. Empty the 4-gallon pail and fill it again from the 9-gallon pail. You now have 1 gallon left in the 9-gallon pail.

b.   Fill the 4-gallon pail and empty it into the 9-gallon pail. Do it again so the 9-gallon pail has 8 gallons. Fill the 4-gallon pail a third time and finish filling the 9-gallon pail—it takes one more gallon to do so. You now have exactly 3 gallons left in the 4-gallon pail.

17. $67 \div 2\frac{1}{2} = 26\frac{4}{5}$; so, 26 packages of seed can be made and there will be $67 - \left(26 \times 2\frac{1}{2}\right) = 2$ ounces of seed leftover.

CHAPTER 2 Fundamental Concepts

SECTION 2.1: Sets

1. **a.** $0 \notin \emptyset$ or $0 \notin \{\,\}$ **b.** $3 \notin B$

3. **a.** {e, l, m, n, t, a, r, y} and $\{x \mid x$ is a letter in the word "elementary"$\}$
   or $\{x \mid x$ is one of these letters: e, l, m, n, t, a, r, y$\}$.

   b. {Spain, Portugal, France, Ireland, United Kingdom (England/Scotland), Western Russia, Germany, Italy, Austria, Switzerland, Belgium, Netherlands, Estonia, Latvia, Denmark, Sweden, Norway, Finland, Poland, Bulgaria, Yugoslavia, The Czech Republic, Slovakia, Romania, Greece, Macedonia, Albania, Croatia, Hungary, Bosnia and Herzegovina, Ukraine, Belarus, Lithuania}.
   Also $\{x \mid x$ is a country in Europe$\}$.

   c. {2, 3, 5, 7, 11, 13, 17, 19, 23, 29, 31, 37, 41, 43, 47, 53, 59, 61, 67, 71, 73, 79, 83, 89, 97}.
   Also $\{x \mid x$ is a prime less than 100$\}$.

   d. The set of fractions between 0 and 1 is infinite.
   $\{x \mid x$ is a fraction between zero and one$\}$.

   e. {name1, name2, name3, etc.}.
   $\{x \mid x$ is a student in this class$\}$.

5. **a.** $\in$; 3 is an element of the set.   **b.** $\subset$; {3} is a subset of the set.

   c. $\in$; {1} is an element of this set of sets.   **d.** $\subset$; {a} is a subset of the set.

   e. $\not\subset$ or $\notin$; {ab} $\not\subset$ is neither a subset nor an element.

   f. $\subset$; the null set is a subset of every set.

7. **a.**

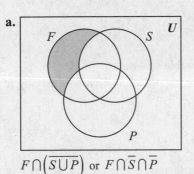

$F \cap \left(\overline{S \cup P}\right)$ or $F \cap \overline{S} \cap \overline{P}$

b.

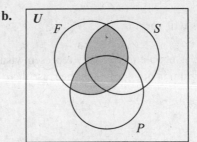

American females who smoke and/or have a health problem.

c.

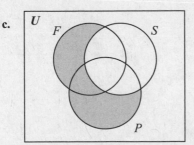

Nonsmokers who either are female or have health problems.

d. $F \cap S$
Females who smoke.

e. $F \cap (S \cap P)$
Males who smoke and have a health problem.

7

9. a.

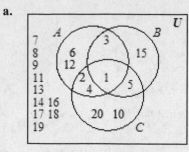

b. All numbers that don't evenly divide 12, 15, or 20; $\overline{A \cup B \cup C}$ or $\overline{A} \cap \overline{B} \cap \overline{C}$

c. All numbers that evenly divide 12 and 20, but not 15; $\overline{B} \cap (A \cap C)$

d. All numbers in U except 1 and 3.

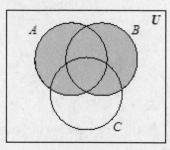

e. All numbers from 1 to 20, except those that divide 12 or 15 evenly.

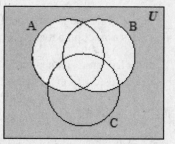

f. Note: This description is ambiguous; it depends on how one interprets "or." $A \cup B$

11. 15 possible committees. Label the members with A, B, C, D, E, and F.
The committees could be: *AB, AC, AD, AE, AF, BC, BD, BE, BF, CD, CE, CF, DE, DF, EF.*

13. Answers will vary.

15. The circles enable us to easily represent visually all the possible subsets.
The diagram is not equivalent because there is no region corresponding to elements that are in all three sets.

17. a. $6 + 8 + 12 + 3 = 29\%$ **b.** $6 + 25 + 15 = 46\%$

c. Those people who agree with his foreign policy and those people who agree with his economic and his social policy.

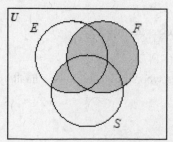

19. a. Construct a Venn diagram. $100 - 11 - 10 - 23 = 56\%$

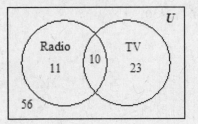

b. $11 + 23 = 34\%$

21. Answers will vary.

23. Answers will vary.

SECTION 2.2: Numeration

1. a.

	Maya	Luli	South American
7		lokep moile tamlip	
8			teyente toazumba
12			caya-ente-cayupa
13		is yaoum moile tamlip	caya-ente-toazumba
15		is yaoum is alapea	
16	uac-lahun	is yaoum moile lokep moile tamop	toazumba-ente-tey
21	hun hunkal	is eln yaoum moile alapea	cajesa-ente-tey
22	ca huncal	is eln yaoum moile tamop	cajesa-ente-cayupa

b. Answers will vary.   **c.** Answers will vary.

3.

		Egyptian	Roman	Babylonian
a.	312	꘠꘠꘡ꓵ∣∣	CCCXII	𒐙 𒌋𒐈
b.	1206	∫꘠꘠ꓵ∣∣∣∣∣∣	MCCIIIIII or MCCVI	𒌋𒌋 𒐋
c.	6000	∫∫∫∫∫∫	MMMMMM	𒐕 𒌍𒌋
d.	10,000	⌒	MMMMMMMMMM	𒐖 𒌍𒐋 𒌍𒌋
e.	123,456	(Egyptian symbols) ꓵꓵꓵꓵꓵꓵ∣∣∣∣∣	Can't do	𒌍𒌍𒐈 𒌋𒐊 𒌍𒌍𒐈𒐈

5. **a.** 26   **b.** 240   **c.** 25   **d.** 450

   e. three thousand four hundred   **f.** 3450

7. **a.**

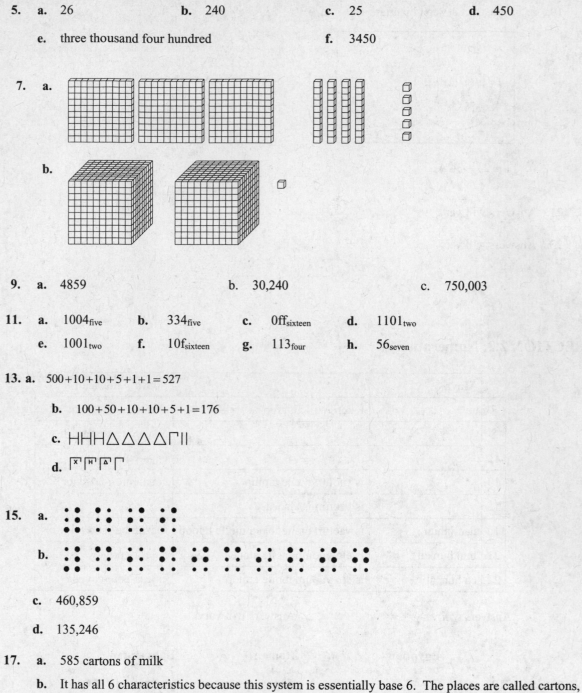

   b.

9. **a.** 4859   **b.** 30,240   **c.** 750,003

11. **a.** 1004_{five} **b.** 334_{five} **c.** $0ff_{sixteen}$ **d.** 1101_{two}

   e. 1001_{two} **f.** $10f_{sixteen}$ **g.** 113_{four} **h.** 56_{seven}

13. a. $500+10+10+5+1+1=527$

   b. $100+50+10+10+5+1=176$

   c. ННН△△△ГII

   d. ⌐x⌐ ⌐н⌐ ⌐△⌐ Г

15. **a.**

   b.

   c. 460,859

   d. 135,246

17. **a.** 585 cartons of milk

   b. It has all 6 characteristics because this system is essentially base 6. The places are called cartons, boxes, crates, flats, and pallets. The value of each place is 6 times that of the previous place.

19. The child does not realize that every ten numbers you need a new prefix. At "twenty-ten" the ones place is filled up, but the child does not realize this. Alternatively, the child does not realize the cycle, so that after nine comes a new prefix.

21. Yes, 5 is the middle number between 0 and 10.

23. We mark our years, in retrospect, with respect to the approximate birth year of Jesus Christ—this is why they are denoted 1996 A.D.; A.D. stands for Anno Domini, Latin for "in the year of our Lord." Because we are marking in retrospect from a fixed point, we call the first hundred years after that point the first century, the second hundred years the second century, and so on. The first hundred years are numbered zero (for the period less than a year after Jesus' birth) through ninety-nine. This continues until we find that the twentieth century is numbered 1900 A.D. through 1999 A.D.

25. Place value is the idea of assigning different *number values* to *digits* depending on their position in a number. This means that the numeral 4 (four) would have a different value in the "ones" place than in the "hundreds" place, because 4 ones are very different from 4 hundreds. (That's why 4 isn't equal to 400.)

27. **a.** 11.57 days   **b.** 11,570 days, or 31.7 years.

29. **a.** 21   **b.** 35   **c.** 55   **d.** 279

    e. 26   **f.** 259   **g.** 51   **h.** 300

    i. 13   **j.** 17   **k.** 153   **l.** 2313

31. base 9

33. $x = 9$

35. This has to do with dimensions. The base 10 long is 2 times the length of the base 5 long. When we go to the next place, we now have a new dimension, so the value will be 2×2 as much. This links to measurement. If we compare two cubes, one of whose sides is double the length of the other, the ratio of lengths of sides is 2:1, the ratio of the surface area is 4:1, the ratio of the volumes is 8:1.

37. **a.**
    b.
    c.

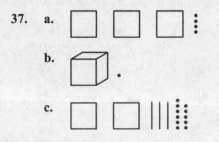

39. Answers will vary.

41. 835

43. $(12 \times 10) + 30,605 = 120 + 30,605 = 30,725$; the correct answer is b.

45. The digit being replaced is in the tens place. So, if the digit 1 is replaced by the digit 5, the number is increased by $(5 \times 10) - (1 \times 10) = 50 - 10 = 40$. The correct answers is b.

CHAPTER 2 REVIEW EXERCISES

1. **a.** $\{x \mid x = 10^n, n = 0, 1, 2, 3, \ldots\}$

　　b. $\{10, 100, 1000, 10,000, \ldots\}$ or $\{10^1, 10^2, 10^3, \ldots\}$

3. **a.** $D \not\subset E$ 　　　　　**b.** $0 \notin \{\,\}$

5. **a.** 　　　　　　　　　　　　　　　　　　　**b.**

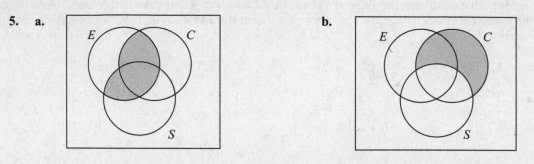

7. The former means the same elements, and the latter means the same number of elements.

9. **a.** 410_{five} 　　　　　　**b.** 1300_{five} 　　　　　　**c.** 1010_{two}

11. $25 \times 2 + 4 \times 5 + 3 = 73$

13. They both have the value of 3 flats, 2 longs, and 1 single. Because base 6 flats and longs have greater value than base 5 flats and longs, the two numbers do not have the same value.

15. The value of the 5th place in base 10 is $10^4 = 10,000$. The value of the 5th place in base 5 is $5^4 = 625$
　　　$10000 \div 625 = 16$.

17. There are several equivalent representations:
　　　$2000 + 60 + 8$
　　　$2 \times 1000 + 0 \times 100 + 6 \times 10 + 8 \times 1$
　　　$2 \times 1000 + 6 \times 10 + 8 \times 1$
　　　$2 \times 10^3 + 0 \times 10^2 + 6 \times 10^1 + 8 \times 10^0$
　　　$2 \times 10^3 + 6 \times 10^1 + 8 \times 10^0$

CHAPTER 3 The Four Fundamental Operations of Arithmetic

SECTION 3.1 Understanding Addition

1. Answers will vary.

3. This pattern can be explained in terms of vectors and with algebraic notation. In vector terms, we know that each horizontal move to the right on the table increases the value by 1, and we know that any vertical move down on the table increases the value by 1. Each move on the diagonals described results from a horizontal move to the right and a vertical move down, and so the value will increase by 2.

   Algebraically, we can find the location in the table of the sum of any two numbers, x and y. We can then use the structure of the addition table to determine the value of other cells in relation to $x+y$. The value of the cell that is diagonally down to the right must be 2 units greater than $x+y$. If we extend the table, we find that the sum of the next term on this diagonal is $x+y+4$, that is $(x+y+2)+2$. Thus, the value of terms on this diagonal increases by 2 each time.

	x	$x+1$
y	$x+y$	$x+y+1$
$y+1$	$x+y+1$	$x+y+2$

5. Answers will vary. One possibility for each is given below.

   a. $47+53=100$ (compatible numbers). Now add 2 more to get 102.

   b. $70+77=147$. Now subtract 1 to get 146.

   c. $56+20=76$. Now take away 1 to get 75.

   d. $575+125=700$. Now add 3 to get 703.

   e. $900+70+13=970+13=983$

   f. $70+30=100$, $100+180=280$, $280+60=340$, $340+(5+5)=350$, $350+13=363$

   g. $387+24=400+11=411$. Then $411+53=464$.

   h. $300+400-5-6=700-11=689$

   i. $295+400-6=695-6=689$

   j. $186+600-2=786-2=784$

7. a. The child is adding one place at a time, beginning with the 1s place but using the break and bridge strategy.

   b. $48+27=48+(20+7)=48+(7+20)=48+(2+5+20)=(48+2)+(5+20)=50+(5+20)=$ $50+(20+5)=(50+20)+5=70+5=75$

   c. $36+4=40$ and then $40+80=120$

   d. $268+2=270$, then $270+300=570$, then $570+40=610$, then $610+5=615$

9. **One explanation:** The digits in the ones place are added first: $6+8=14$. Then the digits in the tens place are added: $3+4=7$. Since the 7 is in the tens place, it has a value of 70. Finally, the two partial sums are added: $14+70=84$.

13

11. Answers will vary.

13. a. The student added the first digits of all the numbers to get $1+1+2+4+1=9$, then the second digits together to get $0+0+0+0=0$, and last the third digits $0+1=1$. The child does not understand the place values of the digits.

b. One possibility is that the student got 20 and then just added the other digits: $2+1+4+9$. The other possibility is that they made a simple addition mistake (this happens frequently with 9s) and had $9+4+2=16$.

c. When adding the ones digits $8+3=11$, the child has placed 11 in the ones column instead of placing 1 in the ones and 1 in the tens column. The child does not understand why numbers greater than 9 are carried to the next column.

15. a. 10100 **b.** 10000 **c.** 154 **d.** bf7

17.

$$4\,2\,0\,4_{\text{six}}$$
$$+\ 3\,5\,5_{\text{six}}$$
$$\overline{5\,0\,0\,3_{\text{six}}}$$

19. Answers will vary. One possibility for each is given below.

a. Leading digit: 1173 $(1000+160=1160; 1160+13=1173)$

b. Leading digit: 16,000. Compatible numbers: $900+100=1000$. Depending on one's short-term memory, one can go even further.

c. Leading digit and rounding: 146,000. Add leading digits: $8+3+2=13$ (representing 130,000). Then round each number to the nearest thousand. And add second leading digit: $9+3+4=16$ (representing 16,000). $130,000+16,000=146,000$.

d. Rounding: $500,000+100,000+700,000=1,300,000$

e. Leading digit, compatible numbers and rounding: 2800. $4+3+1+5+9+3=25$ (representing 2500). Then $73+34=100$; $45+55=100$; $65+43\approx100$. $2500+100+100+100=2800$.

f. Leading digit: 340,000

g. Leading digit and rounding: $2+3+6+3=14$ "hundred thousands"; that is, 1,400,000.

21. a. $<$ goes in the circle. Reasoning: A quick estimation of the 1st three numbers is about 1100. We quickly see that this is less than $624 + 736$.

b. $<$ goes in the circle. Reasoning: Using leading digit, the sum of the first three numbers is 1100 plus less than 100. The sum of the next three numbers is 1000 plus more than 200.

23. a.

$$256$$
$$+\ 588$$
$$\overline{844}$$

b.

$$653$$
$$5246$$
$$+\ 465$$

c.

$$5775$$
$$+\ 7648$$
$$\overline{13423}$$

25. a. The greatest sum is 1593. $(862 + 731; 861 + 732; 832 + 761; 831 + 762)$

   b. The least sum is 405. $(137 + 268; 138 + 267; 167 + 238; 168 + 237)$

27. This is one solution, but there are many more:

$$\begin{array}{r} 372 \\ 168 \\ +459 \\ \hline 999 \end{array}$$

29. Since $100 + 85 = 185 = 105 + 80$, the correct answer is c.

SECTION 3.2 Understanding Subtraction

1. Answers will vary.

3. a. $7 - 5 = 2$ **b.** $7 - 5 = 2$ **c.** $8 - 2 - 2 - 2 - 2 = 0$

5. a. $a - b \neq b - a$, because $b - a = -(a - b)$. For example, let $a = 7$ and $b = 4$. Then, $7 - 4 = 3$, and $4 - 7 = -3$.

   b. $(a - b) - c \neq a - (b - c)$, because $(a - b) - c = a - b - c$ and $a - (b - c) = a - b + c$. For example, $(10 - 4) - 3 \neq 10 - (4 - 3)$, because $6 - 3 \neq 10 - 1$.

7. *One possible way:* $\$145,000 - \$116,000 = (\$145,000 - \$115,000) - \$1,000 = \$30,000 - \$1,000 = \$29,000$

9. *One possible way:* $4132 - 2824$: $32 - 24 = 08$, $41 - 28 = 13$, $4132 - 2824 = 1308$ students

11. a. Samantha is using rectangles to represent tens and dots to represent ones. When she writes out the part to be subtracted, she crosses those pieces off the first list, trading from the tens symbols as needed.

   b. Alice is basically doing the same thing as Samantha, only she writes one of the tens rectangles as a sum giving 10 ($4 + 6$), so she can trade the 6 and leave behind the 4.

13. a.
$904 - 300 = 604$

$604 - 60 = 544$

$544 - 7 = 537$

The method works because
$$904 - 367 = 904 - (300 + 60 + 7)$$
$$= 904 - 300 - 60 - 7$$

   b.
$367 + 7 = 374$

$374 + 30 = 404$

$404 + 500 = 904$

$500 + 30 + 7 = 537$

$904 - 367 = abc \implies 904 = 367 + abc$

First he adds what is needed to get the c digit correct, carrying the extra to the tens column. Then he follows by working on "balancing" the tens and hundreds column.

   c.
$904 - 360 - 544$

$544 - 7 = 537$

This method is like the method in part (a), only the first two steps have been combined.

15. a.   Step 1: Regroup   Step 2: Take away 268

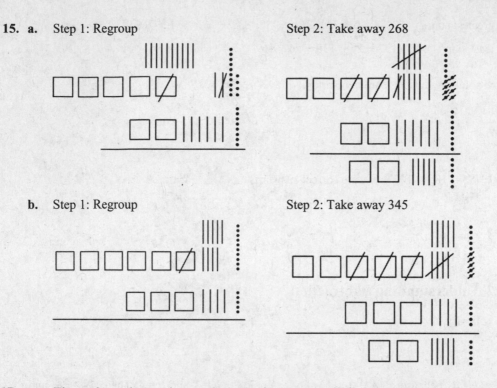

b.   Step 1: Regroup   Step 2: Take away 345

17. a.   The student subtracted, $8-6$. Some students automatically subtract the smaller number from the larger number, even if it changes the order of the numbers in the problem.

b.   The student wrote $6-8=2$, knew that 8 is bigger, and so "borrowed" from the tens place.

c.   The student renamed the 6 in the ones column as 16 before subtracting, but didn't change the 7 in the tens place to 6.

d.   The student probably reversed the numbers in the ones column in order to subtract, $8-0=8$, instead of renaming the 7 in order to subtract, $10-8$. Alternatively, the student could have reasoned that when you have a 0 in the minuend, you just bring down the digit in the subtrahend.

e.   The student changed the zero in the ones place to 10, but didn't rename the 7 as 6 in the tens place.

f.   The student renamed the 7 as 6 and each of the zeros as 10. The zero in the tens place should have been renamed as 9.

g.   The student does not realize he is trading the 10 from the larger place-value column. He know to do $14-9$, but do not change the 6 to 5 to compensate. Thus the next two column subtractions are wrong.

h.   The student is looking at the column $2-9$ and rearranging to actually do $9-2=7$. Since $3-2$ does not need rearranging in the student's eyes, they do $3-2=1$. The student does not understand that there is a difference between $a-b$ and $b-a$.

19. Answers will vary. One possibility for each is given below.

   a. Leading digit: 2100

   b. Round both up to next thousand: $66,000 - 30,000 = 36,000$

   c. 28 to 30 is 2, 30 to 72 is 42, so an estimate is 44,000.

   d. 16 to 26 is 10, to 36 is thus 20, to 43 is thus 27, that is, 27,000.

   e. Think of $413 - 285$. 285 to 300 is 15, to 413 is 113 more. $113 + 15 = 128$. Thus, estimate is 128,000.

   f. $18 + 14 = 32$. So 180,000 to 320,000 is 140,000. Refine the estimate by taking off 4000 to say 136,000.

   g. Rounding: just over 1 million.

   h. Rounding: about 40 million.

21. a. There is more than one solution. b can be any digit, but $a = 0$, $c = 4$, and $d = 2$.

   b. There are seven solutions. In each case, $y = x + 2$. One solution is $x = 2$, $y = 4$.

23. Not reasonable. By using leading digit and looking at the hundreds place, we can quickly see that the answer will be less than 6000.

25. $3020_{five} - 441_{five} = 2024_{five}$

27. Some possibilities: $968 - 734$, $946 - 712$. There are many!

29. The mule was carrying 5 bales of cloth and the horse was carrying 7 bales.

31. The primary reason for the lack of tables is the inverse relationship between the two operations. If you know your addition and multiplication "facts," then you already know the corresponding subtraction and division "facts." For example, if you know that $4 \times 8 = 32$, then you know that $32 \div 8 = 4$, etc. Also, a subtraction or division table is unnecessary—if you know your addition or multiplication facts, you don't need a subtraction or division table. Thus, a subtraction or division table would send a mixed message that this is something new and important. A second reason is that a subtraction or division table would imply more complexity. Since subtraction and division are not commutative, there would be more "facts" to learn.

33. Final odometer reading would be $67,564 + 24 = 67,588$. The total roundtrip distance is $67,588 - 67,324 = 264$ miles. So there are $264 \div 2 = 132$ miles between Fresno and Bakersfield.

35. first day + second day = $547 + 655 = 1202$, and they must sell $1500 - 1202 = 298$ flowers on the third day. The correct answer is a.

37. She must drive $1723 - 849 = 874$ miles. The correct answer is b. [Note: 64% of 4th graders got this correct.]

SECTION 3.3 Understanding Multiplication

1. Answers will vary.

3. 4×3 because we are adding 3 four times.

5. No. Assign numerical values to a, b, and c to show that $a+(b\times c)\neq(a+b)(a+c)$.

7. **a.** The upper and lower halves are symmetrical, which illustrates the commutative property of multiplication.

    b. There are many ways to express the patterns. The most basic is that if you look at the two ends of the diagonal, you find the same number. As you move from the ends to the center of the diagonal, you continue to find matching numbers. Some diagonals have a center number with no partner and some do not. The diagonals with a center number have odd numbers at the ends.

    c. Both products of the diagonals have the same factors. If the factors of the top left term are a and b, the factors of the products of the diagonals are a, b, $a+1$, and $b+1$.

    Using algebraic notation, we can say:

	x	$x+1$
y	yx	$y(x+1)$
$y+1$	$(y+1)x$	$(y+1)(x+1)$

    d-e. See Exercise 9.

9. When nine is multiplied by a number n, the product can be thought of as ten times the number minus the number, $9n = 10n - n$. In this case, $9 \times 7 = (10 \times 7) - 7$. When n is 10 or less, the value of the digit in the tens place of the product is $10(n-1)$, and the value of the digit in the ones place is $9-(n-1)$. Thus the sum of the digits is equal to nine. Since the bent finger represents n, the $n-1$ fingers to the left of n accurately represent the tens place, and the remaining fingers to the right of the bent finger represent the ones place.

11. 900 people

13. **a.**

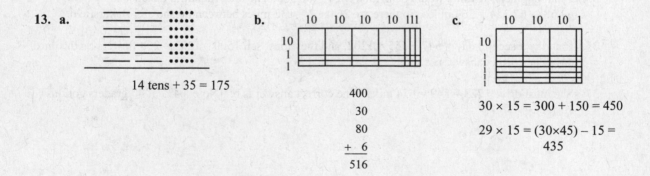

14 tens + 35 = 175

b.

400
30
80
+ 6
─────
516

c.

$30 \times 15 = 300 + 150 = 450$

$29 \times 15 = (30\times45) - 15 = 435$

15.

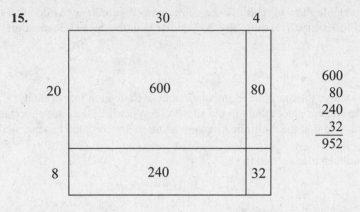

$$
\begin{array}{r}
600 \\
80 \\
240 \\
\underline{32} \\
952
\end{array}
$$

17. a.

$$
\begin{array}{r}
72 \\
\times\ 34 \\
\hline
8 \\
280 \\
60 \\
\underline{2\,1\,0\,0} \\
\hline
2\,4\,4\,8
\end{array}
$$

$\leftarrow 4 \times 2$
$\leftarrow 4 \times 70$
$\leftarrow 30 \times 2$
$\leftarrow 30 \times 70$

This representation is nice because you don't have to do any carrying. Like using FOIL on $(70+2)(30+4)$.

b. This method is almost the same as the method in part (a); the order of multiplication is different.

c.

$$
\begin{array}{r}
72 \\
\times\ 34 \\
\hline
8 \\
28 \\
6 \\
\underline{2\,1} \\
\hline
2\,4\,4\,8
\end{array}
$$

$\leftarrow 4 \times 2$
$\leftarrow 4 \times 7$ (shifted because the 7 is in the tens place)
$\leftarrow 3 \times 2$ (shifted because the 3 is in the tens place)
$\leftarrow 3 \times 7$ (shifted because the 3 and 7 are both in the tens place)

19. a. Set up the numbers as shown at the right.

Find the four partial products and place them in the appropriate places: the ones digit in the lower spot and the tens digit in the higher spot of each cell. Note: If the product of the two numbers is less than 10, you can either leave the top cell blank or place a 0 in that spot.

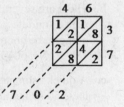

To find the product, find the sum of the numbers on each of the diagonal rows. If the sum of any diagonal row is above ten, "carry" the digit in the tens place to the next place.

b. One way to understand why the lattice method works is to compare it to obtaining the product using expanded form (shown at the right). Looking at both algorithms, you can see the placement of the partial products: 42, 28, 18, and 12. The lattice algorithm somehow enables each digit to be in its proper place.

$$
\begin{array}{r}
40+6 \\
\underline{30+7} \\
42 \\
280 \\
180 \\
1200
\end{array}
$$

You can verify the correct placement of each digit by examining the actual value represented by each partial product. That is, $6 \times 7 = 42$ and the 2 is in the ones place and the 4 is in the 10s place. Similarly, 6×3 represents 6×30, which is 180. In the traditional algorithm, we don't even write the 0 (we just move over). Similarly, in the lattice algorithm, we deal only with "significant" digits: The 8 must be in the tens place and the 1 must be in the hundreds place.

   c. You must know the multiplication table (how to multiply one-digit numbers) and how to add a list of one-digit numbers and "carry" if necessary. You must also be familiar with the ones and tens digit places in base 10.

   d. Answers will vary.

21. Rather than carrying the regroupings above the multiplication (where they are added in immediately following the multiplication step), the student places them in the addition rows (where they are added in during the addition step). She does this so that the multiplication and addition are completely separated.

23. The child put the 8 from 28 below and then also carried the 8.

25. Answers will vary.

27. b is wrong. Estimate it as $20 \times 900 = 18,000$, which is not even close to the answer, 30,102.

29. 3720 miles. Answers on cost will vary depending on price of gas and miles per gallon that the car gets.

31. The value of the largest place contributes more to the final sum than all of the other places combined.

33.   **a.** A novice would say reasonable because $60 \times 80 = 4800$ and thus is under 4800, but more sophisticated students would say that, thinking of the four partial products, it will be clearly less than 4798.

   b. No. This answer essentially takes the smallest and largest of the partial products: 8×4 and 4×3. The answer is also off by one full place: $400 \times 300 = 120,000$—and that error is the larger of the two.

35.   **a.** 72×49. Reasoning 72×49 would be close to $70 \times 50 = 3500$ which is too big.

   b. There are several possible ways to get 5 in the ones place. If we look at partial products and begin with a 'middle possibility' $(90 + 5) \times (60 + 5)$, we see that we have $5400 + 450 + 300 + 25$ which is going to be about 6200. So we can try a bit smaller, $(90 + 3) \times (60 + 5)$. This gives us $5400 + 450 + 180 + 15 = 6045$.

   c. A quick estimate enables to know that the value of the tens place of the second number is 4. Thus we have 50 something times $43 = 2193$. We can get the same answer by dividing 2193 by 43.

   d. A quick glance tells us that we are looking at 80 something times 50 something and the answer has a 1 in the ones place and we need 611 from the partial products. What numbers can give us a 1 in the ones place? We try 83×57 or 87×53. 81×51, and 89×59.

37. A ream of paper is typically 500 sheets of paper.

   a. 600,000 sheets of paper   **b.** $18,000   **c.** $12,000 savings.

39. Answers will vary.

41.   **a.** About 1 per hour, so that means about 24 per day; round to 25 since it is slightly more than 1 per hour. $25 \times 365 = 9125$ per year.

   b. Answers will vary. Some possibilities: How did they find out? Do they test both driver's and passenger's blood for alcohol when there is a fatal accident? Does this imply the drinker is the driver of the car or in the car? Does this count all kinds of traffic accidents: motorcycle, car hits pedestrian?

43. The ones digit of the cube of 156 will be 6. So, a. 3,796,416 might represent the cube of 156.

45. a. This shortcut will work for any two digit numbers that begin with the same number where the sum of the second two digits is 10.

   b. Label the digits of the two numbers to be multiplied: $ab \times ac$. The last two digits of the result come from multiplying $b \times c$. The digits that precede this come from $a \times (a+1)$.

   c.

$$ab \times ac = (10a+b)(10a+c)$$
$$= 100a^2 + 10ab + 10ac + bc$$
$$= 10a(10a+b+c) + bc$$
$$= 10a(10a+10) + bc$$
$$= 100a(a+1) + bc$$

The 100 in the product $100a(a+1)$ simply moves the term two places to the left; thus, the product $a(a+1)$ gives the first part of the answer and bc gives the second part.

47. a. The ones place of the missing multiplier is an 8. Without knowing the product, that is all we can say.

   b. 68×83 will be above 5000 and 78×83 will be above 6000, so the multiplier is 68.

   c. 8×83 and 18×83 will both be less than 2000, so the missing multiplier is either 8 or 18.

49. a. $530 \times 71 = 37{,}630$ **b.** $357 \times 01 = 357$

51. a.

25	17
50	8
100	4
200	2
400	1

$400 + 25 = 425$

b.

48	39
96	19
192	9
384	4
768	2
1536	1

$1536 + 192 + 48 + 96 = 1872$

c.

120	42
240	21
480	10
960	5
1920	2
3840	1

$3840 + 960 + 240 = 5040$

53. a. $(30+4)(20+3) = 600 + 90 + 80 + 12$

$(3x+4)(2x+3) = 6x^2 + 9x + 8x + 12$

If x represents the number 10, then $6x^2 + 9x + 8x + 12 = 600 + 90 + 80 + 12$.

   b. The FOIL algorithm works because it specifies the four partial products.

55. Alike: sum (product) of any row, any column, both diagonals are identical.

Difference: multiplication magic square, don't seem to need all numbers in the square to be different.

57. This is a game to be played by students so there are no answers.

59. Since $200 \times 8 = 1600$ and $30 \times 8 = 240$, the sum is 1840. This means that the digit in the ones place must make a product of $1896 - 1840 = 56$ when multiplied by 8. The number is 7.

SECTION 3.4 Understanding Division

1. 42 jelly beans. The partitioning model.

|||| : |||| : |||| : |||| :

3. **a.** $3 \times 2 = 6$ (repeated addition)

   b. $8 \div 2 = 4$ (repeated subtraction) or $8 \div 4 = 2$ (partitioning)

5. **a.** 30   **b.** 80   **c.** 50   **d.** 30   **e.** 40   **f.** 50

7. **a.-b.**   Answers will vary.

9. There are multiple ways to get answers.

   a. $6 \times 10 = 60$ and $6 \times 5 = 30$ which gives an estimate of 15.

   b. $450/15 = 30$ and $45/15 = 3$, so a quick estimate is 30; a refined estimate is 33

   c. $180/6 = 30$ and $18/6 = 3$, so a quick estimate is 30; a refined estimate is 33

   d.   Cancel the zeros and we have 26/2 which is 13. Here the exact answer is obtained.

   e. $500/25 = 20$ and $75/25 = 3$, so a quick estimate is 20; a refined estimate is 23

   f. $750/25 = 30$ and $100/25 = 4$, so a quick estimate is 30 and the exact answer is 34

   g.   Cancel the zeros and we have 240/5. Since $240/10 = 24$, $240/5 = 48$.

11. The roof is 23 feet 8 inches = 284 inches long and 15 feet 6 inches = 186 inches wide. The area of the roof is 284 inches × 186 inches = 52,824 square inches, which is 366 R120 square feet, or about 367 square feet. Since each can covers 100 square feet and since $367 \div 100 = 3$ R67, I needed to buy 4 cans.

13. $\$8400 \div 12 = \$700$ per month.

15. **a.** 3024_{five} **b.** $203_{\text{five}}\text{r}4_{\text{five}}$ or $\left(203\frac{1}{2}\right)_{\text{five}}$

17. $3240 \div 24 = 675$ cases per week; 675 cases per week × 4 weeks per month = 2700 cases per month for a 5-day week and 4-week month.

19. She must save an additional $\$1234 - \$450 = \$784$. If she saves \$15 per week, it will take $\$784 \div \$15 = 52$ R4 weeks, which rounded up is 53 weeks.

21. 34,000,000 miles $\div$ 16,000 miles per hour = 2125 hours; or $2125 \div 24 = 88$ R13 days, which rounded up is 89 days.

23. Since $75 \div 3 = 25$, the quotient of 75 and a number less than 3 will be greater than 25. So, the correct choice is (a).

25. Not reasonable because $60 \times 60 = 3600$ and if you think about the partial products $(60 + 7) \times (60 + 8)$, you can see that their value is clearly more than 400.

27. Melanie makes $\$10.00 - \$2.50 = \$7.50$ for each tape she sells. To break even, she must sell $\$1200 \div \$7.50 = 160$ tapes.

29. Suppose each glass of juice contains 8 ounces; then each guest will get 16 ounces of juice. Each container has enough juice for 2 people. So, Wei needs to buy $120 \div 2 = 60$ containers of juice. The juice will cost $\$0.89 \times 60 = \$53.40$.

31. One. As I was going to St. Ives . . . everyone else was going the other way!

33. Answers will vary. Assumptions that must be made include the size of the page, the size of the type, the number of columns per page.

35. 3 million crimes per year would mean about 17,000 crimes per school day (dividing by 180 school days per year). If we divide 17,000 by 50 states, that's about 340 crimes per day per state. That still seems high, so the term "crime" must be broad: vandalism, minor assault. Does it mean "crimes reported to police"?

37. Not quite. At this rate, she will burn exactly 370 calories.

39. A carton contains $4 \text{ packs} \times \dfrac{6 \text{ sodas}}{\text{pack}} = 24 \text{ sodas}$. 247 students divided by 24 sodas per carton is 10 remainder 7. Round up to get 11 cartons total.

41. 1961, 30, 48, 20,000, 20

43. a. They will make the trip in less than 30 days.

    b. Exact answer: 5010 miles.

45. Exact answer: $238.19

47. The equation in (c) is wrong, the actual quotient is 78.

49. Francie is correct. There are many ways to arrive at this conclusion. Thinking of partitioning, if you increase the whole (dividend) and decrease the number of groups (divisor), each group will get more. If you increase the whole, then you've got to increase the divisor too.

51. Each child gets 4_{five} gumdrops.

53. a.

a	b	$a + b$	$a \cdot b$
65	66	131	4290
1208	72	1280	86,976

    b.

a	b	$a + b$	$a \div b$
153	9	162	17
19	1	20	19

    c.

a	b	$a - b$	$a \cdot b$
44	24	20	1056

    d.

a	b	$a \cdot b$	$a \div b$
10	2	20	5
34,225	37	1,266,325	925

55. a. $2\overline{)974} = 487$ **b.** $9\overline{)247} = 27.4$

57. a. 123,456 is divisible by 3, 4, 6, and 8

   b. 2,345,678 is not divisible by any of these numbers.

59. a. No

   b. No

   c. Yes

   d. Yes

61. 42,857

63. a. One solution is $66 + 80 - 3$.

   b. One solution is $23 \times 60 + 23 \times 10$.

   c. One solution is $(260 \div 10) \div 2$.

   d. One solution is $25 \times 44 + 1 \times 44$.

   e. One solution is $653 + \mathbf{150} = 803$ and $803 + \mathbf{31} = 834$.

65. a. $[(A + B + C) \times D] / 100$

   b. $(A + B) \times C$

   c. $A + BC - D - E$

67. Since $24 \div 8 = \triangle$, $24 = \triangle \times 8$. The correct choice is d.

69. Since $30 \div 4 = 7$ R2 and $24 \div 5 = 4$ R4, she can make at most 4 necklaces. The correct choice is a.

71. Since the remainder is 5 when 65 is divided by 15, there will be 5 pieces of candy left after she makes the bags. The correct choice is b.

73. The sum is 752; he accidentally multiplied.

CHAPTER 3 REVIEW EXERCISES

1. **a.** $7 + 2$

   b. $7 - 4$

   c. 2×3

   d. $12 \div 3$

3. Answers will vary. This algorithm shows the sum of each place, beginning with the largest place. Once the sum for each place is written down, it is relatively easy to add those numbers mentally to write down the sum.

5. We have simply changed the representation of the minuend from 8 hundreds to 7 hundreds + 9 tens + 10 ones.

7. Answers will vary. At the heart of the matter is that, regardless of the places involved, 10 of something is being exchanged for 1 of something else.

9. The digits being multiplied to get the second row are 37×2. However, this represents 37×20, which is 740. We commonly omit the zero and just write 74; however, because its real value is 740, we must place the 7 and the 4 in the correct places.

11. 541×82. Answers will vary.

13. Answers will vary. The heart of the response is to note that if we are adding, when we take 1 from 17 and give it to 29, it is literally 1. However, if we are multiplying, when we take 1 from 17 and give it to 29, we are really taking one group of 29, rather than just 1.

15. $a = 16$ and $b = 30$ or vice versa.

17. The answer is an infinite sequence beginning with 91 and increasing by 75 each time. That is, 91, 166, 241, 316, and so on.

19. $79 \times 9 \times 11 = 7821$.

21. Answers will vary.

23. **a.** Commutative and associative properties transform the problem into $(36 + 64) + 82$, which equals $100 + 82$.

   b. The distributive property transforms the problem into 20×19.

   c. Representing 1592 as $1600 - 8$, and then using the distributive property of division over subtraction transforms the problem into $\frac{1600}{8} - \frac{8}{8}$.

25. and 27. Answers will vary. One possible response for each is presented here.

25. $345,300 - 216,250 \approx 345,000 - 215,000 = 130,000$

27. $489 \div 19 \approx 500 \div 20 = 25$ miles/gallon

29. When we change 638×42 to 638×40, we are decreasing the answer not by 2 but by 638×2, roughly 1200. Thus, when we round up 638, we need to make up that 1200. If we round 638 to 640, we are increasing by 2×42; if we round 638 to 660, we are increasing by 22×42.

31. The family can travel $18 \times 20 = 360$ miles on one tank of gas. Since the tank is already full, they need to fill up $(1200 - 360) \div 360 =$, which rounded up is 3 times.

33. a. 11330_{five} **b.** 2211_{five} **c.** 3113_{five} **d.** 3214_{five}

35. Base eight

37. a.
$$
\begin{array}{r}
3684 \\
+\ 4248 \\
\hline
7932
\end{array}
$$
b.
$$
\begin{array}{r}
6003 \\
-\ 3284 \\
\hline
2719
\end{array}
$$
c.
$$
\begin{array}{r}
46 \\
\times\ 34 \\
\hline
1564
\end{array}
$$

CHAPTER 4 Extending the Number System

SECTION 4.1 Integers

1. **a.** −218 **b.** −2 **c.** 19 **d.** −78 **e.** 14

   f. −18 **g.** 7 **h.** 2 **i.** 221 **j.** −6

3. −5° Farenheit.

5. $864

7. −$19

9. −40° Celsius

11. **One way:** $-19+(-6)=-25$; $-25+(-22)=-47$; $-47+8=-39$; $-39+(-4)=-43$; $-43+7=-36$; and
    $-36+1=-35$.
    Another way: $\left(-19+(-6)+(-22)+(-4)\right)+(8+7+1)=-51+16=-35$.

13. $2\frac{2}{3}$ hours (2 hours 40 minutes) difference in flight time.

15. Answers will vary. Here are some possibilities.

    a.

1	7	1
5	2	3
1	5	8

    b.

−5	−3	−2
−1	−1	−1
−3	−2	−1

    c.

3	5	−2
8	8	0
−5	−3	−2

    d. The difference in the far right column is $(a-b)-(c-d)$.

    The difference in the far left column is $(a-c)-(b-d)$.

a	b	$a-b$
c	d	$a-c$
$a-b$	$b-d$	$a-b$ $-c+d$

17. Answers will vary.

27

19. a. Always positive. The absolute value of a nonzero expression is always positive.

    b. It depends. If $x > y$, then it will be positive. If $x < y$, then it will be negative.

    c. Always positive. $x^2 + y^2 > xy$

    d. It depends. If $x^2 + 2xy > y^2$, then it will be positive.

21. *One way*: $185 - (-3 + (-2) + 1 + (-6) + 3 + (-2)) = 176$. He weighs 176 pounds now.

23. Since $-2 > -8$ and both numbers are less than 0, the correct answer is b.

SECTION 4.2 Fractions and Rational Numbers

1. Answers will vary. Some possibilities include:

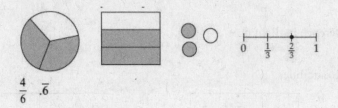

$\dfrac{4}{6}$ $.\overline{6}$

3. larger piece = $\frac{1}{3}$, smaller piece = $\frac{1}{15}$

5. a. $\dfrac{9}{12} = \dfrac{3}{4}$ (1 foot = 12 inches)   **b.** $\dfrac{9}{36} = \dfrac{1}{4}$ (1 yard = 36 inches)

    c. $\dfrac{10}{25} = \dfrac{2}{5}$ **d.** $\dfrac{10}{100} = \dfrac{1}{10}$ (1 dollar = 100 cents)

    e. $\dfrac{8}{32} = \dfrac{1}{4}$ (1 quart = 32 ounces)   **f.** $\dfrac{8}{128} = \dfrac{1}{16}$ (1 gallon = 128 ounces)

7. Answers will vary.

    a. $\dfrac{2}{10}, \dfrac{3}{15}, \dfrac{4}{20}$ **b.** $\dfrac{6}{8}, \dfrac{9}{12}, \dfrac{12}{16}$

    c. $\dfrac{4}{6}, \dfrac{6}{9}, \dfrac{8}{12}$ **d.** $\dfrac{10}{12}, \dfrac{15}{18}, \dfrac{20}{24}$

9. **a.** $\dfrac{25}{40} = \dfrac{25 \div 5}{40 \div 5} = \dfrac{5}{8}$ **b.** $\dfrac{32}{48} = \dfrac{32 \div 16}{48 \div 16} = \dfrac{2}{3}$ **c.** $\dfrac{54}{60} = \dfrac{54 \div 6}{60 \div 6} = \dfrac{9}{10}$

  d. $\dfrac{26}{65} = \dfrac{26 \div 13}{65 \div 13} = \dfrac{2}{5}$ **e.** $\dfrac{168}{216} = \dfrac{168 \div 24}{216 \div 24} = \dfrac{7}{9}$ **f.** $\dfrac{84}{132} = \dfrac{84 \div 12}{132 \div 12} = \dfrac{7}{11}$

  g. $\dfrac{493}{510} = \dfrac{493 \div 17}{510 \div 17} = \dfrac{29}{30}$ **h.** $\dfrac{101,010}{505,050} = \dfrac{101,010 \div 101,010}{505,050 \div 101,010} = \dfrac{1}{5}$

11. **a.** The value of the regions from greatest to least is $\dfrac{1}{2}, \dfrac{1}{3}, \dfrac{1}{6}$.

  b. The value of the regions from greatest to least is $\dfrac{3}{8}, \dfrac{1}{4}, \dfrac{1}{6}, \dfrac{1}{8}, \dfrac{1}{12}$.

  c. Answers will vary.

13.

15. **a.** 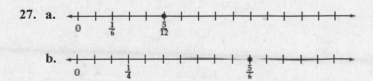 **b.** **c.** 16

17. **a.** 3/7 < 1/2 < 5/8   **b.** 5/6 < 9/10 Each 1 piece away from 1

  c. 2/7 < 1/3 < 4/11   **d.** 7/9 < 15/17 Each 2 pieces away from 1

  e. 2/9 < 1/4 < 5/16   **f.** 3/4 = 75/100 < 79/100

19. If you round both numbers down slightly, you have $\dfrac{35,000}{50,000} \approx \dfrac{7}{10}$.

21. **a.** 5/8

  b. 5/6

  c. 37/158 is between 1/5 and 1/4 . The middle thermometer is in that range.

23. 1/6. Even though there are five sections, they are not equal in size. The shaded portion is 1/3 of 1/2 of the box.

25. This is a valid, though unconventional, response. If you take a whole and divide it into 2 equal parts and then shade in $1\frac{1}{2}$ of them, you have the same area as if you had divided the whole into 4 parts and shaded in 3 of them.

  Alternatively, the ratio $1\frac{1}{2} : 2$ is equivalent to $3 : 4$.

27. **a.**

  b.

29. a. About $\frac{3}{4}$ of $\frac{1}{4}$ of a tank, so $\frac{3}{4} \cdot \frac{1}{4} = \frac{3}{16}$. Given that this is an approximation, one could also say $\frac{2}{3}$ of $\frac{1}{4}$, which is $\frac{1}{6}$.

b. About $\frac{1}{2}$ of a tank plus another $\frac{1}{2}$ of $\frac{1}{4}$ of a tank, so $\frac{1}{2} + \frac{1}{2} \cdot \frac{1}{4} = \frac{1}{2} + \frac{1}{8} = \frac{4}{8} + \frac{1}{8} = \frac{5}{8}$.

31. $\frac{10}{13}$ is closer to 1 than to $\frac{1}{2}$. $\frac{1}{2}$ of 13 is $6\frac{1}{2}$, so $\frac{1}{2} = \frac{6\frac{1}{2}}{13}$. $\frac{10}{13}$ is $\frac{3\frac{1}{2}}{13}$ more than $\frac{1}{2}$ and $\frac{3}{13}$ less than 1.

33. a. 5/15, 3/4, 4/5. 5/15 is less than 1/2; 3/4 and 4/5 are greater than 1/2. 3/4 is 1/4 less than 1; 4/5 is 1/5 less than 1. Since 1/4 is greater than 1/5, 3/4 is less than 4/5.

b. 5/11, 2/3, 6/7. 5/11 is less than 1/2; 2/3 and 6/7 are greater than 1/2. 2/3 is 1/3 less than 1; 6/7 is 1/7 less than 1. Since 1/3 is greater than 1/7, 2/3 is less than 6/7.

c. 3/50, 1/3, 2/5, 5/8, 3/4. 3/50 < 5/50 = 1/10 and 1/10 < 1/3, so 3/50 < 1/3. Since 1/6 is less than 1/5, 2/6 is less than 2/5, so 1/3 is less than 2/5. 5/8 is 1/8 more than 1/2; 3/4 is 1/4 more than 1/2. Since 1/4 is greater than 1/8, 3/4 is greater than 5/8.

d. 3/10, 2/5, 4/7, 5/6, 7/8. 3/10 and 2/5 are the only fractions less than 1/2. 3/10 < 1/3 < 2/5. 4/7 is is closer to ½ than to 1. 5/6 and 7/8 are closer to 1 than to 1/2. Since 5/6 is 1/6 less than 1 and 7/8 is 1/8 less than 1, and 1/6 is greater than 1/8, 5/6 is less than 7/8.

35. a. $\frac{2}{3}$ and $\frac{3}{4}$ are reasonable answers.

b. $\frac{3}{10}$ is closest to the actual fraction, but $\frac{1}{4}$ and $\frac{1}{3}$ are reasonable approximations.

c. $\frac{3}{4}$

d. $\frac{3}{5}$

37. Approximately 1/5 of the trip is left. $24/115 \approx 22/115$, $22 \times 5 = 110$ and $22 \times 6 = 132$, so 1/5 is the best answer.

39. Fuller. There are several ways to justify this. One way: How much larger (multiplicatively) is the denominator? That is, 32 times what = 264 and 58 times what = 402? Mentally, we can determine that $32 \cdot 8 = 256$ and $58 \cdot 7 = 406$.

That is, $\frac{32}{264}$ is slightly less than 1/8 and $\frac{58}{402}$ is slightly greater than 1/7. So $\frac{58}{402}$ is larger.

41. a. There are 48 students (30 girls and 18 boys) in the chorus.

b. 1/4 and 1/5 are both reasonable.

43. a. He was 12 and I was 48. In 6 years, he will be 18 and I will be 54; in 24 years, he will be 36 and I will be 72.

  b. There are many solutions: … (10, 40), (12, 48), (14, 56)…

  c. There are many answers. One is that both numbers are multiples of 10.

45. If the original fraction is less than one, the new fraction is larger. The difference between numerator and denominator becomes less significant as they increase, so they will be proportionately closer together, making for a larger fraction.

47. a. 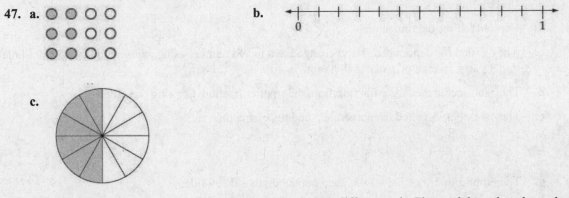   **b.**

  c.

49. The student is trying to compare two fractions that have a different unit. The model needs to have the same unit.

51. a. GCF(12, 30, 75) = 3   **b.** GCF(12, 333, 8415) = 3

53. Each part in Shape R represents $\frac{1}{2}$ of $\frac{1}{6}$, or $\frac{1}{12}$. So, Shape Z represents $\frac{3}{12} = \frac{1}{4}$.

55. The correct answer is a.

57. Since $\frac{1}{2}$ (week 2) $< \frac{5}{8}$ (week 3) $< \frac{3}{4}$ (week 4), the correct answer is c.

SECTION 4.3 Understanding Operations with Fractions

1. **a.** $80\dfrac{17}{24}$ **b.** $-2\dfrac{23}{120}$ **c.** $23\dfrac{13}{20}$ **d.** $191\dfrac{11}{16}$ **e.** $-3\dfrac{7}{12}$

   f. $66\dfrac{7}{24}$ **g.** 99   **h.** 135   **i.** $13\dfrac{1}{2}$ **j.** 1/8   **k.** 3/8

3. **a.** They added the numerators, but instead of first getting a common denominator, they just multiplied the denominators.

   b. They added the denominators.

   c. They added the denominators and changed to a mixed number using a base 10 idea, 13 = 1 whole and 3 parts, instead of 1 whole (8/8) and 5 parts.

   d. They subtracted the larger fraction from the smaller fraction, ignoring the order.

   e. They simply subtracted the numerators and denominators.

   f. $7\dfrac{1}{8}$ should equal $6\dfrac{9}{8}$ not $6\dfrac{11}{8}$.

   g. They multiplied 2×4 and 3×3, then put the digits side by side.

   h. They multiplied both the numerator and the denominator by 5.

   i. They cross-multiplied, one numerator by the other denominator.

   j. They divided the numerators and divided the denominators.

   k. They took the reciprocal of the first fraction instead of the second.

   l. They divided each part separately; $8\div2=4$ and $\dfrac{1}{8}\div\dfrac{1}{4}=\dfrac{1}{2}$.

5. **a.** $\dfrac{23}{6}$ **b.** $\dfrac{23}{4}$

7. Repeated subtraction. Since 5/5 = 1, we repeatedly subtract 5/5 from 13/5 until our remainder is less than 5/5.

9. Diagrams will vary.

a. $\frac{2}{3} \times 2\frac{3}{4}$:

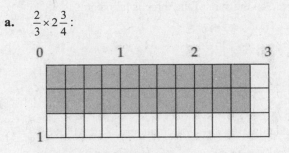

Rearranging the shaded area:

Total area $= 1 + \frac{2}{3} + \frac{2}{12} = 1\frac{5}{6}$

b. $2\frac{2}{3} \times 3\frac{1}{2}$:

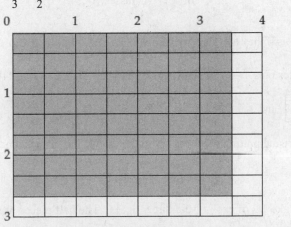

$= 6 + \text{fractional pieces}$

We have 6 whole pieces. Rearranging the fractional pieces:

$= 3 + \frac{2}{6} = 3\frac{1}{3}$

Total area $= 6 + 3\frac{1}{3} = 9\frac{1}{3}$.

11. a. Less than 10. $\frac{5}{8}+\frac{3}{8}=1$ and $\frac{3}{42}<\frac{3}{8}$.

   b.   Less than 2. One possible way: $\frac{3}{4}+\frac{1}{16}<1$, so the sum of the three is less than 2.

   c.   Greater than 2.   Since $\frac{1}{4}>\frac{1}{10}$, $\frac{1}{4}+1\frac{9}{10}>2$.

   d.   Between $5\frac{1}{2}$ and 6. $2\frac{2}{3}<3$. Thus, $8\frac{1}{2}-2\frac{2}{3}>8\frac{1}{2}-3=5\frac{1}{2}$.

   e.   Greater than 20. $8\frac{1}{2}\times 3$ (round up, round down) $=25\frac{1}{2}$. Alternatively, $8\times 2=16$, $8\times\frac{7}{8}=7$; we are already past 20.

   f.   Greater than 1/2. If we double $4\frac{7}{8}$, we will clearly be over 9.

13.   114 teachers.

15.   5/42 of the weight is additives. Let 1 = total weight.

   Subtract the weight of the water and juice: $1-\left(\frac{5}{7}+\frac{1}{6}\right)=1-\left(\frac{30}{42}+\frac{7}{42}\right)=1-\frac{37}{42}=\frac{5}{42}$.

17. a. $213\frac{1}{3}$ ounces left.   **b.** $17\frac{7}{9}$ glasses.

19. a.   The child is adding the fractions as if they are pieces of a pie. The child took $\frac{1}{2}$ from $2\frac{1}{2}$ and added it to $\frac{1}{4}$. This gave 2 wholes plus $1\frac{1}{4}$. The child took apart $2\frac{1}{2}$ into $2+\frac{1}{2}$ and then used the associative property to connect $\frac{3}{4}$ and $\frac{1}{4}$.

   b.   The child is taking fractions apart and putting them back together and using the associative and commutative properties: $2\frac{1}{2}+\frac{3}{4}=\frac{1}{2}+2+\frac{3}{4}=\frac{1}{2}+2\frac{3}{4}=\frac{1}{4}+\frac{1}{4}+2\frac{3}{4}=\frac{1}{4}+3=3\frac{1}{4}$.

21. a.   3/4 is 3 pieces of a 4-piece pie. Take 1/2 of each of the three pieces. The half pieces are each 1/8 of the pie, so three of them would be 3/8.

   b.   3/4 is 3 pieces of a 4-piece pie. Divide two of the three pieces so that you keep one and give one away. Divide the last of the original three pieces in half, keep one and give one away. What you have left is $1/4+1/8=3/8$.

23.   It is not just repeated addition. The most general model of multiplication is that it represents the area of the rectangle formed by the two numbers. If you make a 1×1 square and find 1/4 of that square and then find 1/2 of 1/4 of that square, then you have 1/8 of the square.

25. a.   *One way:* $26\times 11\div 12$. *Another way:* $11\div 12\times 26$.

   b. $26\times 12\div 11$.

27. a. If $4\frac{1}{2} \times 60 = 270$, then $4\frac{1}{2} \times 15 = 67\frac{1}{2}$ and $4\frac{1}{2} \times \frac{1}{2} = 2\frac{1}{4}$. So the answer is $67\frac{1}{2} + 2\frac{1}{4} = 69\frac{3}{4}$.

b. If we double both numbers, we have 9×31. The answer to $4\frac{1}{2} \times 15\frac{1}{2}$ will be $\frac{1}{4}$ of 9×31. If

$9 \times 30 = 270$ and $9 \times 1 = 9$, then $9 \times 31 = 279$. One-half of $279 = 139\frac{1}{2}$ and one-half of $139\frac{1}{2} = 69\frac{3}{4}$.

c. $15 \times 36 = 540$, so $15 \times 18 = 270$ and $15 \times 9 = 135$ and $15 \times 4\frac{1}{2} = 67\frac{1}{2}$. Now we need one-half of $4\frac{1}{2}$,

which is $2\frac{1}{4}$. So $15 \times 4\frac{1}{2} + \frac{1}{2} \times 4\frac{1}{2} = 69\frac{3}{4}$.

29. The conceptual error is that the student does not realize $\frac{a+b}{c+d} \neq \frac{a}{c} + \frac{b}{d}$.

To demonstrate: $\frac{9\frac{1}{4}}{3\frac{3}{4}} = \frac{9+\frac{1}{4}}{3+\frac{3}{4}} \neq \frac{9}{3} + \frac{\frac{1}{4}}{\frac{3}{4}}$, rather, $\frac{9\frac{1}{4}}{3\frac{3}{4}} = \frac{9}{3\frac{3}{4}} + \frac{\frac{1}{4}}{3\frac{3}{4}}$.

31. a. $6 \div \frac{3}{4} = 6 \times \frac{4}{3} = \frac{24}{3} = 8$ guests can be served.

b. Thinking of each pint in 4 parts, there are $6 \times 4 = 24$ parts to split up. Giving 3 to each person, $24 \div 3 = 8$ guests.

c. $\frac{3}{4} \times ? = 6 \implies \frac{3}{4} \times ? = \frac{24}{4} \implies \frac{3}{4} \times \frac{8}{1} = \frac{24}{4}$, thus $\frac{3}{4} \times 8 = 6$.

d. Giving 3/4 to each guest, we'll add $\frac{3}{4} + \frac{3}{4} + \frac{3}{4} + \dots$ until we get 8.

e. We have 6 pints and we'll give away 3/4 pint repeatedly until we run out.

33. a. D   **b.** A   **c.** B   **d.** E

35. 20 packages can be made, with 1 ounce of seeds left over.

37. 13 boxes.

39. The assumption is of monogamy and no sme-sex marriages. That is, the number of married women = the number of married men. Thus, there are more men than women in this community. One solution path: since 2/3 of the women = 1/2 of the men, draw a diagram to represent this fact. (Shown at the right.) Now, all the parts are the same size. Since there are seven parts in the whole, 3/7 of the population is single.

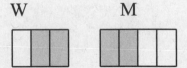

41. It will take 4 pressings to get at least 3/4 of the juice, and 6 pressings to get 9/10 of the juice.

43. a. $48

b. Answers will vary.

45. $\dfrac{2}{9} - \dfrac{1}{5}$

47. $\dfrac{7}{8} \div \dfrac{1}{9}$ if you assumed proper fractions; $\dfrac{8}{2} \div \dfrac{1}{9}$ if you did not.

49. a. 110_{five} **b.** 20_{five}

51. The student is "adding straight across." Either a set, linear, or area model could be used to help with understanding. Using one of these models, the need and process for finding common denominators becomes clear.

52. a. LCM(12, 20) = 60   **b.** LCM(30, 75) = 150

     c. LCM(44, 66) = 132   **d.** LCM(462, 630) = 6930

53 a. LCM(12, 30, 75) = 300   **b.** LCM(20, 30, 40) = 120

55. In 60 seconds, when Ann has run 5 laps and Bob has run 4 laps (60 is the LCM of 12 and 15).

57. There are $\dfrac{3}{3}$ in one cup; so $3 \times \dfrac{1}{4} = \dfrac{3}{4}$. The correct answer is d.

59. He needs 2 × 2 = 4 gallons more to finish painting his fence.

61. The correct answer is b.

63. The given range is not possible for the first bag, since $2 \times 3\dfrac{5}{8} = 7\dfrac{1}{4}$ lb. The second bag (between 10 lb and 11 lb) will hold 3 since $3 \times 3\dfrac{5}{8} = 10\dfrac{7}{8}$ lb. bottles. The third bag (between 14 lb and 15 lb) will hold 4 bottles since $4 \times 3\dfrac{5}{8} = 14\dfrac{1}{2}$ lb.

SECTION 4.4 Beyond Integers and Fractions: Decimals, Exponents, and Real Numbers

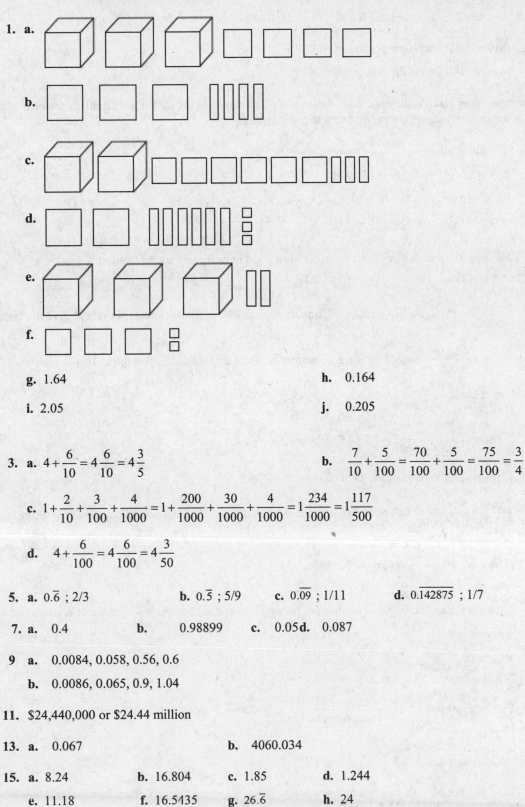

g. 1.64

h. 0.164

i. 2.05

j. 0.205

3. a. $4 + \dfrac{6}{10} = 4\dfrac{6}{10} = 4\dfrac{3}{5}$ **b.** $\dfrac{7}{10} + \dfrac{5}{100} = \dfrac{70}{100} + \dfrac{5}{100} = \dfrac{75}{100} = \dfrac{3}{4}$

c. $1 + \dfrac{2}{10} + \dfrac{3}{100} + \dfrac{4}{1000} = 1 + \dfrac{200}{1000} + \dfrac{30}{1000} + \dfrac{4}{1000} = 1\dfrac{234}{1000} = 1\dfrac{117}{500}$

d. $4 + \dfrac{6}{100} = 4\dfrac{6}{100} = 4\dfrac{3}{50}$

5. a. $0.\overline{6}$; 2/3   **b.** $0.\overline{5}$; 5/9   **c.** $0.\overline{09}$; 1/11   **d.** $0.\overline{142875}$; 1/7

7. a. 0.4   **b.**   0.98899   **c.** 0.05 **d.** 0.087

9 a. 0.0084, 0.058, 0.56, 0.6

b. 0.0086, 0.065, 0.9, 1.04

11. $24,440,000 or $24.44 million

13. a. 0.067   **b.**   4060.034

15. a. 8.24   **b.** 16.804   **c.** 1.85   **d.** 1.244

e. 11.18   **f.** 16.5435   **g.** $26.\overline{6}$ **h.** 24

17. a. 4.1   **b.** 0.64   **c.** 0.027   **d.** -0.135 **e.** 0.03

   f. 42   **g.** 5.6   **h.** 16   **i.** 31.2   **j.** 0.5

   k. 2   **l.** 22.1   **m.** 0.609   **n.** 0.0452

19. a.   Answers will vary. One possibility: 0.9995.

   b.   Answers will vary. One possibility: 3.445.

21. Since the 10ths place is the digit $\div$ 10, the oneths place would be the digit $\div$ 1. But this is the ones place. Thus the ones place and oneths place are the same place.

23.   There are many answers.

   a. $3276 = 36 \times 91$, so $32.76 = 3.6 \times 9.1$.

   b. $476 = 2 \times 238$, so $4.76 = 0.2 \times 23.8$.

   c. $72 \div 2 = 36$, so $0.072 \div 0.2 = 0.36$

25. a.   We want 0.23 of 0.8 m³. The word "of" is a good indication of multiplication:
   $0.23 \times 0.8 = 0.184 \, \text{m}^3$.

   b. $\dfrac{40 \text{ miles}}{\text{gallon}} \times 0.75 \text{ gallons} = 30 \text{ miles}$. Using multiplication, the units cancel properly to give us miles.

   c. $5 \text{ liters} \times \dfrac{0.2 \text{ liters}}{\text{cup}}$ cannot be correct because the units do not cancel. Therefore this is a division

   problem: $5 \text{ liters} \times \dfrac{\text{cup}}{0.2 \text{ liters}} = 5 \div 0.2 = 25 \text{ cups}$.

   d. $7.2 \text{ pounds} \times \dfrac{\text{box}}{0.8 \text{ pounds}} = 7.2 \div 0.8 = 9 \text{ boxes}$

   e. $75 \text{ roses} \div 5 \text{ roses per bouquet} = 15 \text{ bouquets}$

   f. $\dfrac{1 \text{ yard}}{\$15.00} \times \dfrac{1}{0.65 \text{ yards}}$ will not work because the $ units are left in the denominator instead of the numerator.

   Thus, we try $\dfrac{\$15.00}{\text{yard}} \times 0.65 \text{ yard} = \$9.75$.

   g.   We want the answer to be in pounds per person: $5 \text{ pounds} \div 12 \text{ people} = \dfrac{5}{12}$ pound of cookies per person.

   h. $\dfrac{16 \text{ miles}}{\text{sec}} \times 0.85 \text{ sec} = 13.6 \text{ miles}$

   i. $13.9 \text{ meters stretched} \times \dfrac{1 \text{ meter original}}{3.3 \text{ meters stretched}} = 13.9 \div 3.3 \approx 4.2$ meters original length.

27. Answers will vary.

29. 102 is the number of vials that can be completely filled. 0.4 is the fraction of another vial that can be filled. 102 vials will use 127.5 ounces. The remaining 0.5 ounce is 0.4 of a 1.25 ounce vial.

31. $632

33. $70.64

35. a. 1.23456789×10^8 **b.** 3×10^{15} **c.** 5.6×10^{-10} **d.** 3.02×10^{-7}

37. $12,090,000,000,000, which is approximately $40,000 for every person in the United States (based on a population of 300 million).

39. Not much. Assume the average citizen drives 12,500 miles a year and has a car that gets 25 miles per gallon. The person would buy 500 gallons. $0.05/gallon $\times$ 500 gallons = $25.

41. a. The length of each candle.   **b.** Answers will vary.

43. a. Light travels at a speed of 186,000 miles per second. To find out how far it travels in a year, you would do the following calculation:

(186,000 miles/second)(60 seconds/minute)(60 minutes/hour)(24 hours/day)(365.25 days/year)

b. $(14 \times 3356 \times 789) \times 10,000,000,000 = 37,070,376 \times 10,000,000,000 =$ 370,703,760,000,000,000.

45. I guess they figure readers will be confused if they report it accurately, for example, $8\frac{1}{3}$ as 8.3 and $8\frac{2}{3}$ as 8.7.

47. $21/60 = 0.35$, so it is 7.35 minutes.

49. a. 3.75

b. 6.67

c. 7:12 P.M.

d. 8 A.M.

51. a. Diagram will be equivalent to 1/5.

b. A0.A.

53. a. 4.8

b. 4.05

55. The correct answer is e.

57. The correct answer is d.

CHAPTER 4 REVIEW EXERCISES

1.   **a.**   −4   **b.**   1/4

3.   −2

5.   The first symbol represents an operation. The second symbol signifies the value of the number. To mix them up is to become sloppy, which is not a good habit to get into. The number sentence given reads, "Negative 3 minus 4 is equal to negative 7."

7.   Because $\dfrac{5}{6}=\dfrac{35}{42}$ and $\dfrac{6}{7}=\dfrac{36}{42}$, you can go to 84ths and see that $\dfrac{71}{84}$ is between the two.

9.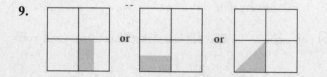

11.   Answers will vary. Figure should have four dots.

13.   When we divide 1 by 2 in base 5, we get $0.2222..._5$.

15.   There are many ways to express the reason. One is to say that, by definition, the numerator of a fraction means how many "equal" pieces one has. If the denominators are not identical, then the numerators represent pieces of different size.

17.   We can solve it without the algorithm by finding $7\dfrac{3}{4}$, $2\dfrac{1}{3}$ times – that is,

$$7\frac{3}{4}+7\frac{3}{4}+\frac{1}{3}\left(6\frac{3}{4}+1\right)=7\frac{3}{4}+7\frac{3}{4}+2\frac{1}{4}+\frac{1}{3}=18\frac{1}{12}$$

19.   She can make 15 cakes and will have $1\dfrac{1}{4}$ cups of flour left over.

21.   Answers will vary.

23.   **a.**   0.625   **b.**   0.004   **c.**   6/25

25.   −0.54   0.045   0.454   5/11

27. $4,306,000,000

29.   **a.**   0.112 cubic meters. We multiply because we have part of the tank: fraction as operator.

      b.   53 cups. We divide because the problem is repeated subtraction – repeatedly taking away 0.15 liter.8 boxes.

      c. $6.06. This can be seen as division as a proportion: $\dfrac{4.85}{0.8}=\dfrac{x}{1}$.

CHAPTER 5 Proportional Reasoning

SECTION 5.1 Ratio and Proportion

Notes: (1) All of these problems can be solved in more than one way; some of the alternative solution paths are noted. (2) This is not an algebra course; most of these problems can be solved with proportions, but there are no problems for which solving a proportion is the only solution path.

1. 143.75 calories.

3. 165 dentists.

5. a. $750   b. $1250   c. $977.08

7. 860,000 people.

9. Approximately $51\frac{1}{2}$ feet.

11. Yes.

13. $11\frac{3}{4}$ inches. Approximate answer.

15. a. 48 cents

    b. 96 cents

    c. $4.41

17. Answers will vary.

19. a. 0.15 lb

    b. 14 oz or 0.875 lb

    c. 49 million

21. Amy's red blood cell count is high.

23. 10 questions

25. 3.5 feet.

27. 37 handshakes per minute, which is one handshake in less than 2 seconds.

29. 2520 students.

31. $1\frac{1}{2}$ hours more.

33. There are more prison inmates. Assuming a U.S. population of 300,000,000, there are about 800,000 physicians and about 1,500,000 prison inmates.

35. The ratio between boys and girls becomes greater, because the ratio of boys to girls added to the class, 1:1, is greater than the original ratio, 3:8.

37. a. The proportions of the ingredients vary with the quantity, due to the cooking process.

   b. 8 cups liquid, 5 cups cereal, 1/2 teaspoon salt. Answers may vary.

39. 199/325 = 61.2% and 5/8 = 62.5% These percentages are close enough to say that the advertisement is accurate. Also, 199/325 ≈ 200/325, which simplifies to 8/13, but 5/8 sounds better than 8/13 for advertising purposes.

41. The person who gets paid twice a month will have a larger paycheck.

43. If you assume that the truck slowed her down by 20 miles per hour and Sonja claims that she would have arrived at class 20 minutes earlier, then she followed the truck for $6\frac{2}{3}$ miles. It seems unlikely that she would not have been able to pass the truck in this distance, unless she was driving on a narrow, winding road.

45. a. Answers will vary. For example, there are more than 100,000 more Black males in prison than white males.

   b. Answers will vary. For example, the rate of incarceration for black males is more than 7 times the rate of incarceration of white males.

   c. Answers will vary. For example, from the raw numbers, the numbers for white females and black females are close. However, the rates for black females is almost 5 times the rate for white females. The disparity or inequity becomes more visible with rates.

47. a. Answers will vary. For example, the numbers in South Africa are slightly more than the number for India.

   b. Answers will vary. For example, while the numbers for South Africa and India are similar, the rate in South Africa is more than 20 times the rate in India.

   c. Answers will vary.

49. Additive decline 5.2 vs 8.5. Multiplicative decline 48% vs 38%.

51. 1 orange has the same amount as 2 eggs, so 6 oranges has the same amount of calcium as 12 eggs. The correct answer is d.

SECTION 5.2 Percents

1. **a.** 18
   b. 19.55

   c. 1211.8
   d. 54.4

   e. 0.8625
   f. 240

   g. 80%
   h. 85.7%

   i. 8.52%
   j. 26.1%

   k. 18.75%
   l. 145.8%

   m. 46.35
   n. 480

   o. 7.192
   p. 70

   q. 32.4
   r. 43.65

   s. 1185.6
   t. 160

3. 35%. Using compatible numbers, $23 \times 3 = 69$, so 23 is just over 33%.

5. 700 patients. One way to estimate is to use guess–check–revise. 30% of 1000 is 300, so try smaller. Using this method, you can get the actual answer.

7. **a.** 131%. Estimating: 42,000 to 84,000 is 100% increase. Answer is more than 100%.

   b. 26%. Estimating: Using compatible numbers, 15,000/60,000 = 1/4.

   c. 43%. Estimating: Increase is about 31,000, or just under 1/2.

   d. 132%. Estimating: Similar strategy as in (a).

9. Approximately 6%. The baby lost 1/2 pound. $\frac{1}{2} \div 8 = \frac{1}{16} \approx \frac{6}{100}$. Adult would have lost approximately 10 pounds. If you see $6\% \approx 1/16$, then $1/16 \times 160 = 10$.

11. About 12.8 million. 12,812,800

13. Approximately 41% of eligible voters aged 18–24 voted in 2008.

15. $\dfrac{\$1760}{x} = \dfrac{80}{100} \Rightarrow 80x = 17,600 \Rightarrow x = \$2200$

17. $0.055x = \$53 \Rightarrow x = \$53 \div 0.055 = \$963.64$

19. After the first year, she gets a raise of $0.55 \times \$28,200 = \$1551$. So her new salary is $29,751. After the second year, she gets a raise of $0.04 \times \$29,751 = \$1190.04$. So after two years her salary is $30,941.04.

21. $0.044 \times \$21,430,000 = \$942,920$. The new budget is $22,372,920, which rounds to $22,373,000.

23. About 1.2 miles.

25. The "whole" each year is not the same.

Let's take someone making $40,000. If they get a 6% raise per year, their salary after four years would be $50,499.08. If they got 12% in the first year and then 4% for the next three years, their salary after 4 years would be $50,393.91. However, and here is where math is not simple, the total four year's salary under the first plan would be $185,483.70 compared to $190,241.60 under the second plan. So after four years, they would have more total salary under the second plan but their base salary would be slightly higher under the first plan.

At a simpler level, think of $40,000 with a 50% raise the first year and then no raise the second year vs. a 25% raise each year. In the first plan, your salary jumps to $60,000 and then stays at $60,000 the second year. Under the second plan, your salary jumps to $50,000 the first year and then it goes up 25% of $50,000 the second year and so it goes up to $62,500.

27. a. The high school principals make 91% more than the teachers; the junior high school principals make 79% more than the teachers; and the elementary school principals make 68% more than the teachers.

   b. The junior high school principals make 6% less than the high school principals. The elementary school principals make 12% less than the high school principals. The teachers make 48% less than the high school principals.

29. a. 5 out of 1000, or 1 out of 200, children will have a severe reaction.

   b. Answers will vary. Possible questions include the following: What are the risks of not having the child vaccinated? How does the reaction compare with the illness itself? Are some children more likely than others to have a serious reaction?

31. 7% grade means that a hill changes 7 vertical feet for every 100 horizontal feet, or 70 vertical feet for every 1000 horizontal feet, or a proportionate amount of change. The percent grade gives drivers an indication of the steepness of a hill.

33. a. The monthly grocery bill would increase $0.072 \times \$325 = \$23.40$. The new average monthly grocery bill is $348.40. This amounts to $\$348.40 \times 12 = \$4180.80$ per year.

   b. The raise was less than the rate of inflation. $\$1200/\$23,400 \times 100 \approx 5\%$ raise.

35. If the cost of the item is less than $100, take $10 off. If the cost of the item is greater than $100, take 10% off.

37. a. $10.29

   b. $20,580

   c. This assumes that she works 40 hours a week and 50 weeks a year.

   d. She would be making $10.33 an hour, which is about $80 more a year.

39. $121,885.98

41. $\$1000 \times 0.06 \times t = \$1000 \Rightarrow 60t = \$1000 \Rightarrow t = \$1000 \div 60 = 16\frac{2}{3}$ years.

43. $2\frac{1}{2}$ hours

45. a. Using leading digit to get the total number of cars, we get about 140 million and 13/140 is approximately 1/10 or 10%.

   b. 9.8%

   c. These data are probably available from the Registry of Motor Vehicles, which every state has. I would guess these data are pretty reliable.

47. Impossible to say without knowing how many people total attended each game.

49. a. $48\,\text{in} \div 3\frac{3}{4} = 12$ with 3in remaining

   b. $\frac{3}{48} \times 100 = 6.25\%$ waste

   c. 6.25% of $40,000 = $2,500 wasted

   d. A 45-inch coil would result in no waste (although there are certainly other correct answers)

   e. Both coils result in less waste than the 48-inch coil.

   f. Answers will vary. It is essential to explain that since $48 \div 3.75 = 12.8$, the 0.8 represents 8/10 of the 3.75 in that is needed.

51. $5375 + 0.15(\$5375) = \$6181.25$. The correct answer is e.

CHAPTER 5 REVIEW EXERCISES

1. $69.29

3. The ratio will increase.

5. 559 miles

7. The birth rate is a ratio. Thus, to keep the ratio the same, if the population increases, the number of births also needs to increase. If the rate is down, it means that the number of births did not increase at the same rate as the population did.

9. Still need to raise 15% of the goal, which is $240,000.

11. $22,373,000

13. The task will be complete in one more hour.

15. $1,575,000

17. 4.6% of the world's population lived in the U.S. in 2008.

19. $\frac{1470}{1840} = \frac{x}{100} \Rightarrow 1840x = 147,000 \Rightarrow x \approx 80\%$; her take-home pay is about 80% of her gross pay.

21. I had punched 0.0055 instead of 0.055.

23. If the original price is greater than $100.

25. Technically it comes to $305,300, but the more reasonable answer is $300,000, because 115% is an average.

27. An increase of 71%.

CHAPTER 6 Algebraic Thinking

SECTION 6.1 Understanding Patterns, Relations, and Functions

1. Answers will vary.

3. Examine the first four triangular numbers to determine a pattern:
   1st: 1 = 1
   2nd: 3 = 1 + 2
   3rd: 6 = 1 + 2 + 3
   4th: 10 = 1 + 2 + 3 + 4
   Each triangular number n is the sum of the first n natural numbers.

   a. Since $1 + 2 + 3 + 4 + 5 = 15$, there are 15 dots in the fifth triangular number.

   b. Since $1 + 2 + 3 + 4 + 5 + 6 + 7 + 8 + 9 + 10 + 11 + 12 = 78$, there are 78 dots in the 12th triangular number.

   c. The nth triangular number is the sum of the first n natural numbers:

   $$1 + 2 + 3 + \cdots + (n-2) + (n-1) + n = \frac{n(n+1)}{2}.$$

   So the number of dots in the nth triangular number is given by $\frac{n(n+1)}{2}$.

5. a. Each bus can hold 36 passengers: 34 students and 2 adults. So, $232 \div 34 = 6$ R28. Therefore, 7 buses are needed for 232 students.

   b.

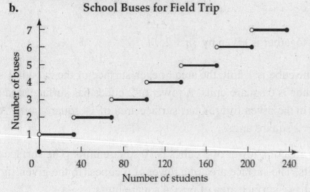

7. Examine the given figures to determine the pattern.

   1st: $5 = 1 \cdot 5 = 1 \cdot (2 \cdot 1 + 3)$
   2nd: $14 = 2 \cdot 7 = 2 \cdot (2 \cdot 2 + 3)$
   3rd: $27 = 3 \cdot 9 = 3 \cdot (2 \cdot 3 + 3)$
   4th: $44 = 4 \cdot 11 = 4 \cdot (2 \cdot 4 + 3)$
   $\vdots$
   nth: $n \cdot (2 \cdot n + 3)$

   The number of squares in the nth figure in this pattern is determined by multiplying n by the odd natural number determined by $2n + 3$. So, the number of squares in the nth figure in the pattern is given by $n(2n + 3)$, or $2n^2 + 3n$.

9. Examine the given figures to determine the pattern.

1st: $9 = 3 \cdot 3 = 3 \cdot (4 \cdot 1 - 1)$
2nd: $21 = 3 \cdot 7 = 3 \cdot (4 \cdot 2 - 1)$
3rd: $33 = 3 \cdot 11 = 3 \cdot (4 \cdot 3 - 1)$
$\vdots$
nth: $3 \cdot (4n - 1)$

So the number of squares in the nth figure is given by $3(4n - 1)$, or $12n - 3$.

11. a. $T_n^2 - T_{n-1}^2$

b. Yes, it is a function. Justifications will vary.

13. Determine the perimeter of the given shapes to find a pattern.

Number of pentagons	1	2	3	4
Perimeter	5	8	11	14

For each shape, n = number of pentagons.

Perimeter of 1st shape ($n = 1$): $5 = 3 \cdot 1 + 2$
Perimeter of 2nd shape ($n = 2$): $8 = 3 \cdot 2 + 2$
Perimeter of 3rd shape ($n = 3$): $11 = 3 \cdot 3 + 2$
Perimeter of 4th shape ($n = 4$): $14 = 3 \cdot 4 + 2$
$\vdots$
Perimeter of nth shape ($n = n$): $3 \cdot n + 2$

The perimeter of n pentagons joined together is given by $3n + 2$.

15. a. Since the length of each side of the cube is 1 unit, the area of each surface of the cube is 1 square unit. So the surface area of one cube is 6 square units. A tower of 2 cubes has surface area of 10 square units. A stack of 3 cubes (in the given figure), has surface area of 14 square units. A tower of n cubes has a surface area of $4n + 2$ square units.

b. (See figures below.) The surface area of a tower of 2 cubes is 10 square units. The surface area of a tower of 4 cubes is 16 square units. The surface area of a tower of 6 cubes (in the given figure) is 22 square units. A tower of n cubes has a surface area of $6n + 4$ square units.

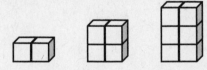

17. a.

Number of hours	0	1	2	3	4	5	6	7	8
Number of people	2	4	8	16	32	64	128	256	512

It will take 8 hours until all 450 students in the school hear the rumor.

b.

Number of hours	0	1	2	3	4	5
Number of people	2	6	18	54	162	486

It will take 5 hours until all 450 students in the school hear the rumor.

19. Answers will vary.

21. a. Answers will vary. Some patterns may be that the y values are increasing by 3 each time, or that they are 3(1), 3(2), 3(3), etc.

   b. When x is 100, y is 300.

   c. The y value is 3 times the x value.

   d. $f(x) = 3x$

23. a. $4^2 + 5^2 + 20^2 = 21^2$

   b. $1^3 + 2^3 + 3^3 + 4^3 + 5^3 = 15^2$

   c. $16 + 17 + 18 + 19 + 20 = 21 + 22 + 23 + 24$

SECTION 6.2 Representing and Analyzing Mathematical Situations and Structures Using Algebraic Symbols

1. Let x be the number. Then:

$$\frac{3x+6}{3} - x = \frac{3(x+2)}{3} - x = x + 2 - x = 2$$

3. Let x be the number. Then:

$$\frac{2x+10}{2} - 5 = \frac{2(x+5)}{2} - 5 = x + 5 - 5 = x$$

5. Answers will vary.

7. The models are rectangles for even numbers, and the rectangles with the "one left over" added on for odd numbers. If we take an even rectangle and multiply it an even or an odd number of times, it will still be a rectangle (even). If we take an odd rectangle with the one left over and multiply it an odd number of time, there will still be one left over.

   Algebraically,

   even times even $= 2k \times 2n = 2(k \times 2n) =$ even (2 times a number)
   odd times even $= (2k + 1)(2n) = 2(2k + 1)(n) =$ even (2 times a number)
   odd times odd $= (2k + 1)(2n + 1) = 4kn + 2k + 2n + 1 =$ odd (1 left over)

9. **a.**   The balance scale below shows the first setup.

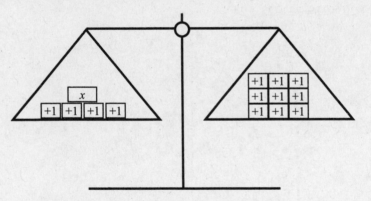

   To solve for x, we remove (or subtract) four +1s from each side to keep the scale balanced and we are left with x on one side and eight +1s on the other side. So, $x = 8$.

b. The balance scale below shows the first setup.

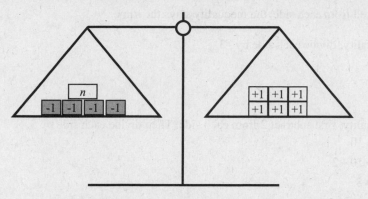

To solve for *n*, we add four +1s to each side to keep the scale balanced, leaves *n* on one side and ten +1s on the other side. So, *n* = 10.

c.

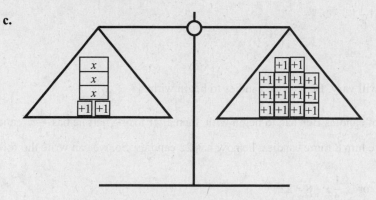

First we remove (or subtract) two +1s from each side. Then, we can see that each *x* would equal four +1s. So, *x* = 4.

d.

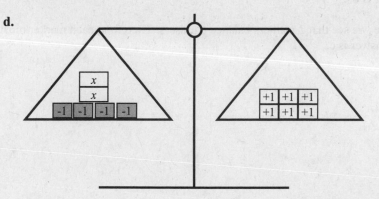

First, we add four +1s to each side, which would leave two *x*'s on one side and ten +1s on the other side. Then we can see that each *x* would equal five +1s. So, *x* = 5.

11. a. Yes. When we subtract 4 from each side, we get −2 < 1.

   b. When 4 is subtracted from each side, the inequality stays the same.

13. a. To solve the inequality, divide each side by −3.

$$\frac{-3x}{-3} < \frac{9}{-3}$$
$$x > -3$$

   b. To solve the inequality, first subtract 2 from each side. Then divide each side by 5.

$$5x + 2 > 10$$
$$5x + 2 - 2 > 10 - 2$$
$$5x > 8$$
$$\frac{5x}{5} > \frac{8}{5}$$
$$x > 1\frac{3}{5}$$

15. Answers will vary.

17. a. Solution methods will vary. Ben has 28 candies to begin with.

   b. If x is the number of candies Ben had to begin with, then after losing half he has $\frac{1}{2}x$ candies left.

   After his Mom gave him 8 more candies, he now has 22 candies. So, we can write the following equation:

$$\frac{x}{2} + 8 = 22 \quad \text{or} \quad \frac{1}{2}x + 8 = 22$$

19. In one day, Lee delivers ☐ newspapers; in two days he delivers ☐ + ☐, or 2 × ☐ newspapers; and so on. The correct answer is b.

21. From the balance scale, we see that 2 markers balance 6 erasers. Therefore, each marker must balance 3 erasers. The correct answer is c.

SECTION 6.3   Using Mathematical Models to Represent and Understand Quantitative Relationships

1. Answers will vary.

3. **a.** Values in the table may vary.

Number of People	Fee
10	190
20	230
30	270
40	310
50	350

b. Let C represent the fee (or cost) for catering a banquet and P represent the number of people. Then, the fee for catering a banquet can be expressed by the equation: $C = 150 + 4P$.

c.

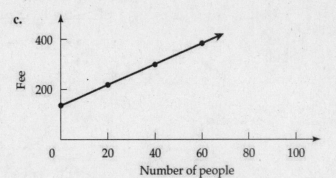

d. Using the equation from part b, we have:

$$462 = 150 + 4P$$
$$312 = 4P$$
$$78 = P$$

So, 78 people attended the banquet.

e. Answers will vary.

5. **a.** Solution methods will vary. One possible solution method is to write an equation. If x is the number of checks you write in one month, then the first plan can be expressed with the equation $C = 2 + 0.15x$, where C is the amount charged for writing x checks. You need to determine when the first plan is a better option than the second plan; that is, when you are charged less than $5 per month. Solve the equation:

$$5 = 2 + 0.15x$$
$$3 = 0.15x$$
$$20 = x$$

The charge for both types of checking accounts is the same if you write 20 checks per month. If you write fewer than 20 checks per month, choose the first plan. If you write more than 20 checks per month, choose the second plan.

b. Methods will vary.

c. Both are functions. For both plans, there is only one possible charge for the number of checks you write in one month.

7.   Substituting 45 into the formula, we get: $H = 2.3(45) + 61.4 = 103.5 + 61.4 = 164.9$ cm, or about 1.65 m.

9.   Thunder travels at approximately $\dfrac{750}{3600}$, or 0.208 miles per second. In 5 seconds the sound of thunder travels $5(0.208) = 1.04$ miles (or about 1 mile), which is how far away the lightning struck.

11. Since 3 books cost $12, one book costs $4. We can express the cost of a book (B) plus a pen (P) plus a notepad (N) costs as the equation $B + P + N = \$14$. We can use the fact that a book plus a notepad costs $11—that is, $B + N = \$11$—to replace $B + N$ in this equation, which gives: $P + \$11 = \$14$. Therefore, $P = \$3$. Finally, replacing B with $4 and solving the equation $\$4 + N = \$11$, we see that $N = \$7$.

13. From the third equation, the sum of an oval and a triangle is 9, and we can replace $\bigcirc + \triangleright$ in the first and second equations with 9 and solve to find the value of one oval and the value of the rhombus.

$$\diamondsuit + 9 = 15; \quad \diamondsuit = 6 \qquad\qquad \bigcirc + 9 = 14; \quad \bigcirc = 5$$

Since the value of an oval is 5, we can see from the third equation that the value of a triangle is:

$$5 + \triangleright = 9; \quad \triangleright = 4$$

15. a.

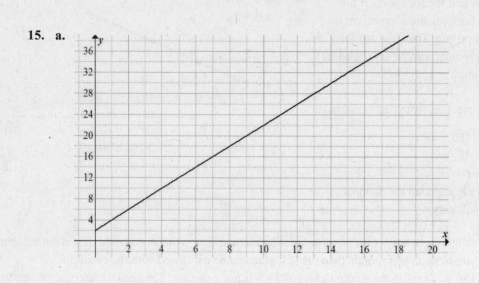

     b.   A 10-mile taxi ride would cost $22.

17. a.

     b.   We can write two equations: $A + T = 1200$ and $A = 600 + T$. Replacing A in the first equation with $600 + T$ and solving, we have:

$$(600 + T) + T = 1200$$
$$600 + 2T = 1200$$
$$2T = 600$$
$$T = 300$$

Thomas earned $300 and Adam earned $600 + $300 = $900.

SECTION 6.4 Analyzing Change in Various Contexts

1. a. When it us 2 years old, the value of the machine is $3400.

    b. In 1 year, the machine decrease $800 in value.

    c. Since the value decreases in value $800 per year, the value of the copy machine will be zero in

    $5000 \div $800 = $6\frac{1}{4}$ years.

3. The first graph.

5. Answers will vary. Sample answer:

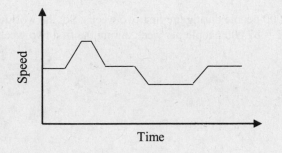

7. Answers will vary.

9. a.–f. Answers will vary.

11. a. The ramp rises 1 foot for every 12-foot increase horizontally.

    b. The slope of the ramp must be equivalent to $\frac{1}{12}$. So, $\frac{1}{12} = \frac{6}{x}$; $x = 6(12) = 72$. The ramp will start 72 feet from the building.

13. a. Geometric; to get the next term, we multiply by 2.

    b. Arithmetic; to get the next term, we add 4.

    c. Neither; these are the square numbers.

15. Answers will vary.

17. Answers will vary. Sample answer:

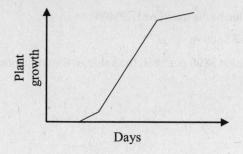

No, it is not a constant rate of change, because the slopes of the lines change.

19. There was an increase of 174,000 people during the first two weeks. So, the worldwide population increased at a rate of 174,000/2 = 87,000 people per week during the first two weeks of 2014.

CHAPTER 6 REVIEW EXERCISES

1. Let S represent the number of student tickets sold and A represent the number of adult tickets sold. The amount raised was \$1342 and 458 tickets were sold. Using guess, check, and revise, we have:

    If $S = 200$, then $A = 258$: \$2(200) + \$5(258) = \$400 + \$1290 = \$1690

    If $S = 250$, then $A = 208$: \$2(250) + \$5(208) = \$500 + \$1040 = \$1540

    If $S = 300$, then $A = 158$: \$2(300) + \$5(158) = \$600 + \$790 = \$1390

    Note that, with an increase of 50 student tickets, the total amount raised decreases by \$150. This means that the total amount raise decreases by \$3 for each 1 ticket increase in the number of student tickets sold. Since \$1390 is \$48 more than the actual amount raised, we divide by \$3 to determine how many more student tickets were sold: \$48 ÷ \$3 = 16. So, if 16 more student tickets were sold the amount raised would be \$2(316) + \$5(142) = \$632 + \$710 = \$1342.

3. Butterflies flap their wings $\dfrac{12}{1 \text{ seconds}} \times \dfrac{60 \text{ seconds}}{1 \text{ minute}} \times \dfrac{60 \text{ minutes}}{1 \text{ hour}} = 43,200$ times in 1 hour.

5. **a.** Each term of the sequence is 1 more than a multiple of 6: $7 = 6(1) + 1$; $13 = 6(2) + 1$; $19 = 6(3) + 1$; $25 = 6(4) + 1$. The next term is $6(5) + 1 = 31$; the 20th term is $6(20) + 1 = 121$; and the nth term is $6n + 1$.

    b. Each term is a power of 2: $1 = 2^0$; $2 = 2^1$; $4 = 2^2$; $8 = 2^3$; $16 = 2^4$. The next term is $2^5 = 32$; the 20th term is $2^{19} = 524,288$; and the nth term is $2^{(n-1)}$.

    c. Each term is 1 less than a perfect cube: $2 = 3^1 - 1$; $8 = 3^2 - 1$; $26 = 3^3 - 1$; $80 = 3^4 - 1$. The next term is $3^5 - 1 = 242$; the 20th term is $3^{20} - 1 = 3,486,784,400$; the nth term is $3^n - 1$.

    d. Each term is 1 less than 6 times a power of 2: $5 = 6 \cdot 2^0 - 1$; $11 = 6 \cdot 2^1 - 1$; $23 = 6 \cdot 2^2 - 1$; $47 = 6 \cdot 2^3 - 1$. The next term is $6 \cdot 2^4 - 1 = 95$; the 20th term is $6 \cdot 2^{19} - 1 = 3,145,727$; the nth term is $6 \cdot 2^{(n-1)} - 1$.

7. **a.**

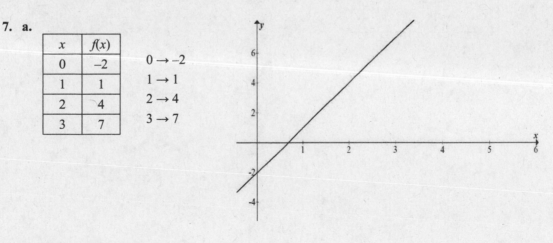

x	$f(x)$
0	–2
1	1
2	4
3	7

$0 \rightarrow -2$

$1 \rightarrow 1$

$2 \rightarrow 4$

$3 \rightarrow 7$

   b. $f(6) = 3(6) - 2 = 16$

9. Examine the given figures to determine the pattern.

1st: $7 = 2 + 5 = 2 \cdot 1 + 5$
2nd: $9 = 4 + 5 = 2 \cdot 2 + 5$
3rd: $11 = 6 + 5 = 2 \cdot 3 + 5$
4th: $13 = 8 + 5 = 2 \cdot 4 + 5$
$\vdots$
nth: $2 \cdot n + 5$

It will take $2n + 5$ squares to make the nth figure in the pattern.

11. Look for a pattern.

1st: $2 = 1 + 1 = 1^2 + 1$
2nd: $6 = 4 + 1 = 2^2 + 1$
3rd: $12 = 9 + 1 = 3^2 + 1$
4th: $20 = 16 + 1 = 4^2 + 1$
5th: $30 = 25 + 5 = 5^2 + 5$
$\vdots$
nth: $n^2 + n$

a. Based on the above pattern, the 10th rectangular number is $10^2 + 10 = 110$.

b. The nth rectangular number is $n^2 + n$ of $n(n + 1)$.

13. If he averages more than 400 minutes per month, then he should pick the second plan. Otherwise, he should pick the first plan.

15.

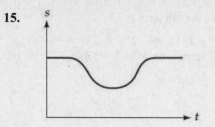

CHAPTER 7 Uncertainty: Data and Chance

SECTION 7.1 The Process of Collecting and Analyzing Data

Note: In cases where students are asked to explain the graph or to describe questions about reliability, validity, etc., there are many possible valid ways to answer those questions. Because of space limitations, only one response is given here. However, it should be interpreted as "one of many possible valid responses," as opposed to the right response or even the best response.

1. Answers will vary. Possibilities are given here:

    a. How many times can a third grader dribble a ball in one minute – only one hand can be used, the ball must visibly bounce off the floor each time, it can only touch the floor (i.e., not bounce off the wall).

    b. How long can you hop on one foot – person can select one foot but cannot alternate from one foot to the other, everyone must hop with arms at sides or hop with arms out straight, all hop with same footwear the other foot should be in same position, e.g., just off floor or by the knee, etc., can't touch anything with hands, for each hop the foot must be visibly off the ground.

    c. How many concerts have you attended in the past year – can include free or paying, just focused on music, a street festival counts as one concert.

    d. How much time do you study in a week – include all time spent on your courses outside class, reading, studying, working on projects. Round to the nearest hour.

3.
```
    x  x
    x  x
    x  x
  x x x x
  x x x x   x
  x x x x  x x  x x            x       x x    x
  1 1 2 2 2 2 2 2 2 2 2 2 3 3 3 3 3 3 3 3 3 3 4 4 4 4 4 4 4
  8 9 0 1 2 3 4 5 6 7 8 9 0 1 2 3 4 5 6 7 8 9 0 1 2 3 4 5 6
```

    a. There were 27 students whose ages range from 18 – 46. The cluster, from 18 – 21, contains 18 students or 2/3 of the class. Only 4 students are older than 27.

    b. 24. That feels like the center of gravity, as if all these x's were on a see-saw.

    c. mean = 24, median = 21, modes = 19 and 21

5. a. Mean is 123, median is 122, and mode is 121.

    b. They do not tell us anything about the shape of the data—the range, the spread, gaps, clusters, or outliers.

    c.
```
                              x
                              x
                            x x
                          x x x        x
          x             x x x x  x x  x x
          x     x  x    x x x x x x x x x x x   x x   x    x
   107 109 111 113 115 117 119 121 123 125 127 129 131 133 135 137 139
```

    d. The number of raisins range from 107 to 138; most lie between 117 and 129.

59

e. It does not tell you the average.

f.

100–109	2
110–119	6
120–129	25
130–139	4

g.

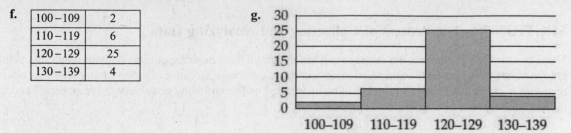

h. The histogram tells us that most of the boxes had between 120 and 129 raisins.

i.

100–104	0
105–109	2
110–114	2
115–119	4
120–124	16
125–129	9
130–134	2
135–139	2

j.

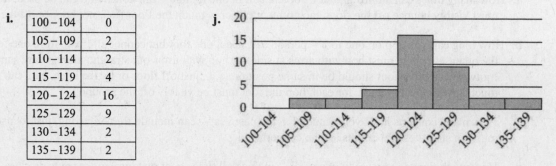

k. The histogram tells us that most of the boxes had between 120 and 129 raisins.

l. They are alike in that they both show a spike in the 120s. They are different in that the second histogram shows that the largest concentration is in the 120-124 range and that, other than the 120s, the number of raisins in each interval is relatively constant.

m. Answers will vary.

n. They are packaged by weight, not by number of raisins.

7. a. Answers will vary.   **b.** The mean is 4.2 years; the median is 4 years.

9. 93

11. Let x be the number that is removed. $\dfrac{4(7)+x}{5} = 6 \implies 28 + x = 30 \implies x = 2$.

13. a. Mean is 72.75 and median is 77.5.

b. Delete one below and one above the median.

c. Delete two below the median.

d. Delete two scores whose sum is greater than 146.

e. Delete one score above and one below the median; the sum of the scores must be greater than 146.

15. a.

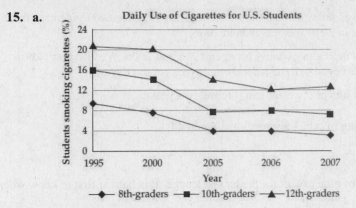

b. For all grade levels, the percentage of students that are smoking cigarettes has decreased over the last 12 years.

c. 8th grade; decrease of $\approx 74\%$; 10th grade ; decrease of $\approx 56\%$; 12th grade ; decrease of $\approx 52\%$

d. Answers will vary.

17. a.

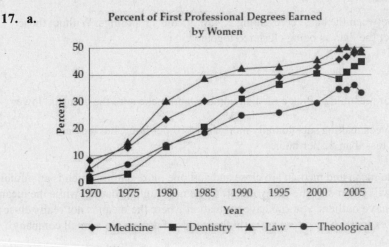

b. Answers will vary.

c. The percentages of women receiving degrees in these fields have risen dramatically in the past 36 years. By 2006 almost 5 out of every 10 degrees in medicine and law went to women, and about 3 in 10 theological degrees went to women.

Answers for 19-23 will vary. Possible answers are given.

19. a. Almost $\frac{3}{4}$ of Americans drink at least one cup of coffee a day.

b. There is no information about how they got the data. Did they ask people of different ages, incomes, ethnicities?

c. No problem with the graph.

21. **a.**   The percentage of employees given a day off for Martin Luther King Day seemed to increase from 1986 to 2006, but the percentage in 2010 was less than in 2006.

 b.   We don't know that the numbers increased steadily between 1986 and 2006. We also don't know what kinds of employees were surveyed. All the federal employees get the day off.

 c.   The x-axis is not labeled, though it is pretty clear that it represents years.

23. **a.**   About $\frac{2}{3}$ of Americans dying of AIDS are between 25 and 44.

 b.   What year is this for?

 c.   Each wedge is a different shade of blue except for the largest wedge. It is hard at first to know where Under 15 is.

25. **a.**   The mode is New Hampshire. The mode is the datum that occurs the most.

 b.   There is no mean. Although you can, and computer programs will, compute the numbers, the number 4.6 is meaningless.

27. **a.**   Mean is $44/12 = 3.7$, median is 2.

 b.   There should be spaces between the bars representing 2, 4, 6, 9, and 11 siblings. Without those spaces, one could interpret the data as being clustered together.

29. Answers will vary.

30. Answers will vary. One line of reasoning: Pick a median number, and make sure the numbers lower

31. There are at most a few employees making substantially more than $10 per hour, but they are outnumbered by those making less than $7 per hour.

33. Think of a set of data where the mean and median are close and add one more datum which is an outlier on the high end. The median will either stay the same or shift to the next highest datum, while the mean will jump appreciably. If you have outliers, you can have a situation where the mean is not really close to the center of the data. See the example below for wages (in thousands of dollars) in a small company:

    12, 12, 12, 15, 15, 15, 18, 18, 18, 18, 18, 18, 20, 20, 20, 30, 30, 30, 75, 100

In this case, the median is 18,000. The mean is 26,000 and ¾ of the employees make less than 26,000. That is, ¾ of the data are below average.

35. Answers will vary.

37. Rates allow us to compare different wholes. For example, say there were 20 murders last year in a city of 800,000 and 30 murders in a city of 1,800,000. If we just show the raw data, and the reader did not know the populations of the city, one could conclude that the second city was more dangerous. However, if we use 100,000 as our unit, we would say that the first city has a murder rate of 2.5 per 100,000 people and the second city has a murder rate of 1.7 per 100,000 people.

39. What is crucial is that there is a whole. Many data can be in percentages but there is no whole. For example, see problem 23.

41. a.

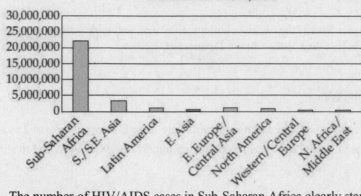

Estimated Number of HIV/AIDS
Cases in the World, 2013

b. The number of HIV/AIDS cases in Sub-Saharan Africa clearly stands out on this graph.

c. Answers will vary.

43. a. Answers will vary.

b.

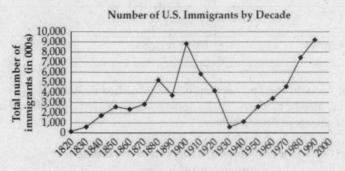

The first graph makes it appear that there has been a rapid increase in the number of immigrants in the last few decades.

The second graph gives a more accurate impression that the <u>rate</u> of increase is actually not as high as it has been in the past.

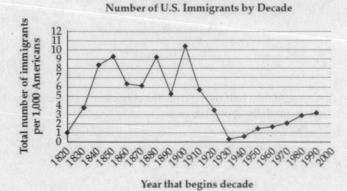

c.-d. Answers will vary.

45. a. The number of AIDS diagnoses for Blacks is equal to that for all other groups combined.

b. Since the data came from the Centers for Disease Control and Prevention, the data are probably accurate. Hoowever, as with problem 44. I would wonder about illegal immigrants and homeless people. Are the numbers for thise populations underreported?

c. A bar graph makes sense. A circle graph would make it easier to realize that Blacks represent $\frac{1}{2}$ of the total number of cases.

47. a. In Germany and the Unites States, the overwhelming majority of seatwork time is on practicing procedures, whereas it is less than $\frac{1}{2}$ the time in Japanese classrooms.

   b. Same as for problem 46.

   c. The choice of a bar graph is fine.

49. a. About $\frac{5}{6}$ of all immigrants to the United States in 1900 were from Europe, as opposed to about $\frac{1}{6}$ in 2000. Latin Americns represented less than 2% of the immigrants in 1900 compared to almost $\frac{1}{2}$ in 2000.

   b.

	1900		2000
Europe	84.9%	Europe	15.3%
Latin America	1.3%	Latin America	51.0%
Asia	1.2%	Asia	25.5%
Other regions	12.6%	Other regions	8.1%

   c. With the table, you lose the breakdown of the sections of Latin America in 2000 unless you make another table; but that would also be cumbersome for some readers. With the graph, you can more visually see the tremendous change in demographics.

51. a. If you enclose each column of x's in a line plot with a bar, you have a histogram.

   b. If you put a dot at the center of the top of each bar and connect the dots, you have a line graph.

53. Answers will vary.

55. Between 1970 and 1980

57. The correct answer is c.

59. I would use the median. In both cases, more than half of the data are very close to the median. In the first case, four of the five data values are substantially above the mean.

SECTION 7.2 Going Beyond the Basics

1. **a.** It tells us the range: 52 – 81 and that there is a cluster in the mid to high 60s. We can say that about ½ the class has a pulse rate between 65 and 69.

```
                      x x x
 x           xx    xx  xxxxx        xx        x
 5 5 5 5 5 5 5 6 6 6 6 6 6 6 6 6 6 7 7 7 7 7 7 7 7 7 8 8
 2 3 4 5 6 7 8 9 0 1 2 3 4 5 6 7 8 9 0 1 2 3 4 5 6 7 8 9 0 1
```

b. It tells us the range and that ost of the pulses are in the 60s.

```
5 | 289
6 | 236778899
7 | 145
8 | 1
```

c. It tells us the range, that the median is 67, and that about half the class's pulse is between 62 and 70.

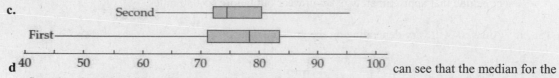

```
 5 5 5 5 5 5 5 6 6 6 6 6 6 6 6 6 6 7 7 7 7 7 7 7 7 7 8 8
 2 3 4 5 6 7 8 9 0 1 2 3 4 5 6 7 8 9 0 1 2 3 4 5 6 7 8 9 0 1
```

d. I would have everyone sitting down. I would make sure that everyone got their pulse in the same way, e.g., finger on the wrist. I would say "1, 2, 3, start" and then "stop" after 30 seconds. Some people don't have the best concentration and so getting the number for 30 seconds and then doubling it would be fine.

3. **a.**

```
               4 | 5
               5 |
         32 |  6 | 47
  9764444310 |  7 | 02256
         821 |  8 | 012348
           6 |  9 | 48
```

b. The first class has a much larger range (45 to 98) compared to (62 to 96). The first class has about the same number of scores in the 70s as in the 80s, while the second class has a large cluster in the 70s.

c.

```
Second ————————[   |   ]————————————————
First  ——————————[    |    ]——————————————
   40       50       60       70       80       90      100
```

d can see that the median for the first class is higher (78 vs 74). We can also see that the middle half of the scores are similar (about 71 to 83 in the first class and about 72 to 80 in the second class).

e. Answers will vary.

5. The second class had a median of 79, as compared with 77 in the first class. The mean for the first classe is 80.8. The mean for the second class is 80. The second class was bimodal at 77 and 85, whereas the mode for the first class was 76. The second class had a smaller range, 65 to 96 compared with 58 to 100 in the first class.

7. a. Grouped frequency bar graph, histogram, or circle graph is appropriate. Also appropriate are box-and-whisker and line plots.

b. Since the distribution is skewed, the mean, median, and mode are not convergent. The median is 21. The data are bimodal at 19 and 21. The mean is 23.9.

c. Data is skewed to the right.

Range: 18-46
Clusters: 18-21
Biggest gap: between 27 and 37
Outliers: 37, 42, 43, 46
Standard deviation: 8.0

9. a.

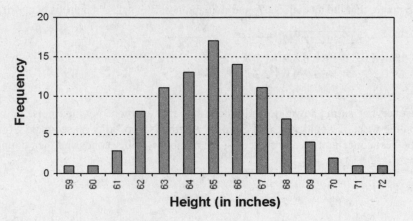

b. Mean: 65.15; standard deviation 2.43.

c. 70%

11. 61; 71. About 1% are less than 5 feet tall.

13. a. 50 tires.

b. Without *z*-scores, one can approximate the area under the curve. By various means, one can conclude that approximately 6% will wear out before 55,000 miles.

15. a. Answers will vary depending on the grouped data.

b.

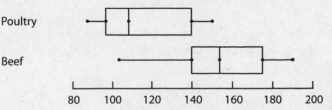

c. Answers will vary.

d. The relationship between calories and sodium content is pretty strong. That is, in general, the more the calories the more the sodium.

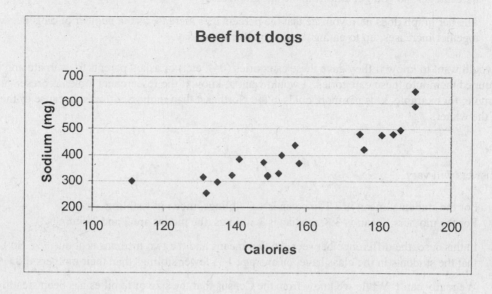

17. a. Positive correlation   **b.** No correlation

c. Negative correlation   **d.** Positive correlation

e. Positive correlation   **f.** Positive correlation

g. Positive correlation   **h.** Positive correlation

19. and 21. Answers will vary. Below I describe interesting aspects for each set of data.

19. a. Whom did you survey? I would not say that over half of the families I know eat dinner together 5 or more days a week. Does it count if only part of the family is there? Were these data gathered from two-parent families or from one- and two-parent families?

b. How often does your family eat dinner together in an average week during the school year? (I would give them the categories in the table below or I would ask for a specific number. For example, a response of "2 or 3" would create problems in comparing to the data given.)

c.

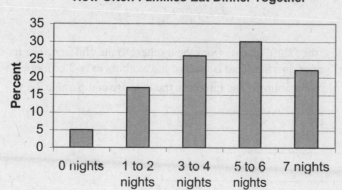

d. A circle graph would have been okay here, but there is a numerical progression (from none to 1 or 2 to 3 or 4 to 5 or 6 to 7), and it is easier to follow this progressiopn with a bar graph. One main advantage of a circle graph is that it gives you the part of the whole; in this case, the data are in percentages, so you get that from the bar graph also.

e. The bar graph does help you see that the percentages increase as the number of days per week eating together increases, up to eating dinner together every day.

21. I would want to know if they gave these categories (0-4, etc.) or asked parents to estimate and then grouped them into these categories. I would want to know if the researchers took a representative sample; for example, kids get more colds in the Northeast than in the Southwest because of the severity of the winters.

23. 76.7

25. Answers will vary.

27. a. For the students: mean is 2.1, median is 1.5 siblings, mode is 1 sibling.
For the mothers: mean is 3.8, median is 3 siblings, the modes are 2 and 3 siblings.

b. In this case, the difference between the two means and the two medians is about 1 ½. So I might say that the students in the class have, on average 1 ½ fewer siblings than their mothers did.

c. We really can't. While we know from the Census that the size of families has been steadily decreasing, the demographics (characteristics) of the students in the class is by no means representative of the overall population.

d. The line plots show that the data for the students is more tightly clustered around 0 to 2 and then steadily decreases after that. The 11 is clearly an outlier.

```
    X X
    X X   X
   X X X   X
  X X X X X     X   X     X
  0 1 2 3 4 5 6 7 8 9 10 11
        parents
    X
    X
    X X
   X X X
   X X X X
   X X X X X X       X
   0 1 2 3 4 5 6 7 8 9 10 11
        students
```

e. The boxplots nicely show the "shift" in data from the mothers to the children. The mothers' median is equal to the students' upper quartile. That is, half of the mothers have 3 or more siblings; only $\frac{1}{4}$ of the students have 3 or more siblings. The ranges in the data are comparable.

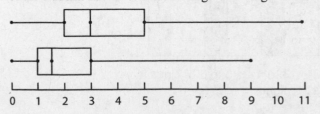

   f. The standard deviations are 2.1 and 2.8, respectively. The mean absolute deviations are 2.11 and 2.17, respectively.

   g. We would need to know the ages of each student and the student's mother.

   h. It would tell us if the size of a student's family is related to the size of the mother's family. That is, do students with small families tend to have mothers from small families and students with larger families tend to have mothers from larger families?

   i. Answers will vary.

29. a. Make sure all of the intervals have the same number of years in them.
    b. Answers will vary.

    c. Answers will vary.

    d. The graph from part (a) clearly shows that there are many brand-new teachers in the district—which is not evident from looking at the average.

31. a. The football payrolls range from about $62 to $120 million with a median of about $82 million. The middle half of the teams have payrolls between $77 and $91 million. There appears to be a cluster of payrolls between $77 and $81 million. The baseball payrolls range from about $30 to $210 million with a median of about $66 million. The middle half of the teams have payrolls between $48 and $87 million. The longest whisker is longer than the other whisker and boxes combined which indicates that there are probably gaps between $87 and $210 million and that $210 million might be an outlier.

    b. The football distribution is probably slightly skewed to the right. The baseball distribution is strongly skewed to the right.

    c. The baseball payrolls have a range (about $123 million) that is more than the range of the football payrolls (about $58 million). About $\frac{1}{2}$ of the baseball teams have a payroll smaller than all of the football teams.

33. a. A cluster

    b. A high probability of gaps, or in the case of whiskers, the probability of outliers.

    c.
    Normal distribution is symmetric.

    d.
    Skewed to the right means more clustered at the left and more spread out at the right.

35. Without *z*-scores, one can approximate the area under the curve. By various means, one can conclude that approximately 35%, or about 700 tires, will wear out before 45,000 miles.

37. Her math score was the highest in terms of standard deviations above the mean.

39. Answers will vary.

41. There is a mild positive correlation.

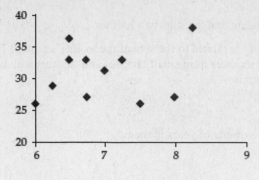

43. a. If we are going to make generalizations about the population called "teenage boys," the driving records of the boys surveyed in these places is likely to be worse than if we had a representative sample.

   b. If we are going to make generalizations about the population called "citizens," then parents is not a random sample. Even if we are going to make generalizations about the population called "parents," then parents at a PTA meeting is still not representative. Only a fraction of parents attend PTA meetings.

   c. If we are going to make generalizations about the population called "residents," this will not be a representative sample: Not all people have telephones, and certain people are home less than others, for example, young single people.

   d. You would probably get different results if you asked their opinion on December 23!

45. Her mean speed is about 17 miles per hour.

47. Julie is right. Although 148.12 grade points / 46 credits = 3.22 GPA, one's total grade points is either a whole number or a mixed number.

SECTION 7.3 Concepts Related to Chance

1. 1/3

3. 0.14 or 1 in 7 ; 0.77

5. a. 0.1   **b.** 0.3   **c.** 0.6   **d.** 0.4   **e.** 0.9

7. a. $\frac{1}{4}$ **b.** $\frac{3}{8}$

9. a. $\frac{1}{36}$ **b.** $\frac{18}{36} = \frac{1}{2}$ **c.** $\frac{15}{36} = \frac{5}{12}$ **d.** $\frac{6}{36} = \frac{1}{6}$

11. 2/3 of a chance of landing in Room A.

13. 2/9

15. 1/6

17. No, it is not a fair game.

19. The expected value from the spinner is \$1.625. Take the \$2 and run. Over the course of 52 weeks, the spinner would yield \$84.50, while taking \$2 per week would yield \$104.

21. 2/3

23. \$375.

25. a. It is fair.

   b. It is not fair. Give 1 point to player A if the number is even and 3 points to player B if the number is odd.

   c. It is fair.

27. a. $\frac{24}{48} = \frac{1}{2}$ **b.** $\frac{1}{2} \times \frac{23}{47} = \frac{23}{94}$

29. 1/12

31. 53/80

33. $\frac{53}{64} \approx 0.83$

35. a. $5/16 = 0.3125$

    b. $35/128 \approx 0.27$

    c. $63/256 \approx 0.25$

    d. Answers will vary.

    e. $\dbinom{50}{25} \div 2^{50} \approx 11\%$

37. The most likely sum is 5. The probability of rolling a 5 is $4/16 = 1/4$.

39. Answers will vary. $P(3 \text{ doubles in a row}) = \dfrac{1}{6} \times \dfrac{1}{6} \times \dfrac{1}{6} = \dfrac{1}{216} \approx 0.00463$.

41. Answers will vary. Here is one set of possibilities: One die must have the same number on all its faces. The other die could have 1, 1, 3, 3, 5, 5, or three of one odd number and three of another.

43. The probabilities of winning are: player 1: 1/9, player 2: 2/9, and player 3: 6/9
Thus, give each player this number of points when they win: player 1: 6 points, player 2: 3 points, player 3: 1 point.

45. No.

47. No. We need to remember the law of large numbers. If you play many times, you are more likely to win about ¼ of the times. But with a small sample size, the unlikely is possible though not probable, like rolling doubles 4 times in a row, for example.

49. Facts: There are at least one red, one blue, and one green ball and there are at least three different colors of balls in the bag. Inferences will vary - there are probably more red than blue (or green) balls; there are probably at least twice as many red as blue (or green) balls; there are probably less than 10 colors, etc.

51. Answers will vary.

SECTION 7.4 Counting and Chance

1. a. 1/52 **b.** 1/13 **c.** 3/13

2. $6! = 6 \times 5 \times 4 \times 3 \times 2 \times 1 = 720$ ways.

3. a. $4 \times 3 \times 2 \times 1 = 4! = 24$

b. 16, determined by making the arrangements

4. a. 24 possibilities **b.** 96 possibilities

5. 10,000

6. $P(\text{face card}) = 12/52 = 3/13$.

$P(\text{picking three face cards in a row with replacement}) = (3/13)^3 = 27/2197$.

7. $3/51 = 1/17$

8. $_{10}P_4 = 5040$ ways, because each way associates a certain flag with a particular position on the flagpole.

9. $_{12}C_3 = 220$ ways.

10. a. $_9C_5 = 126$ possible starting lineups.

b. $_8C_4 = 70$ lineups.

11. If the flavors were scooped in any order, there would be $_9C_3 = 84$ possibilities. If you specified the order of the flavors, there would be $_9P_3 = 504$ possibilities.

12. The probability is $\dfrac{1}{5040}$.

13. $P(4 \text{ of a kind}) = 0.0002$. $P(3 \text{ of a kind}) = 0.0211$. $P(2 \text{ of a kind}) = 0.423$.

14. $7 \times 6 \times 5 = 210$ ways.

15. 1/120

16. 729 combinations.

17. The probability that each child will get the popsicle he or she wants is $\dfrac{26}{27}$.

18. 17/24.

19. a. 21 work combinations

b. 7. (SM, MT, TW, WT, TF, FS, SS)

c. 6 choices

20. If you can choose among 5 toppings, there are $2^5 = 32$ possible pizzas. Thus, there are $32 \times 32 = 1024$ possible combinations. The number in the commercial (1,048,576) is equivalent to 2^{20}. It is possible that the persons who designed the commercial may have reasoned that since there are 2^5 combinations for the first pizza, the number of possible combinations for the two pizzas was $2^5 \cdot 2^4$, and then reasoned that $2^5 \cdot 2^4 = 2^{20}$.

21. 39,916,800 possible words from *mathematics*, 11!. Assuming you use all the letters.

22. Answer will be n! divided by the factorial of each letter represented. Thus KANSAS would have $(6!/(2!2!))$ arrangements, that is 180 arrangements.

23. n! can be written as follows: $n(n-1)(n-2)...(n-r+1)(n-r)!$ When $_nP_r$ is expressed as a fraction, $(n-r)!$ in the numerator and the denominator cancel each other, leaving $n(n-1)(n-2)...(n-r+1)$, which is the other expression for $_nP_r$.

24. $_nP_n$ means that n positions are to be filled by n people or objects, $_nP_n = n!$.

25. Answers will vary.

26. Answers will vary. Be sure to consider the likelihood of students having the same first name or the same last name.

CHAPTER 7 REVIEW EXERCISES

1. a.

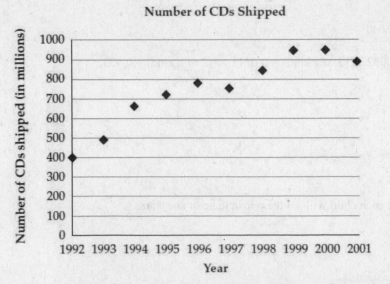

b. The number of CDs shipped rose steadily between 1992 and 2001, with a dip in 1997 and 2001. The number of units shipped in 2001 was more than double the number shipped 9 years earlier.

3. a.

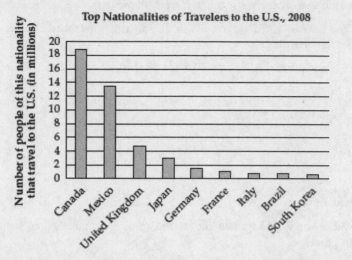

b. Answers will vary.

5. Responses for these questions will vary.

7. a. One possible graph is below.

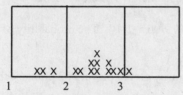

b. The mean is 23.7; the median is 25; and the mode is 25.

c. The number of drops recorded varied from 15 to 30. Over half of the data were between 21 and 27 drops.

9. a.

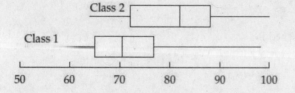

The box plot gives a quick snapshot. It tells us that the first class had a signigficantly higher range; that the second class did better overall - it had a higher median; and that 3/4 of the second class is above 70, compared to only 1/2 of the first class.

b.

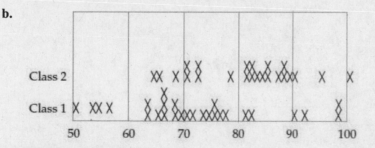

The line plot also lets us see the range; it also lets us see the clusters. The first class is relatively spread out; the second class has a cluster in the 80s.

    c. *Box:* pros - quick snapshot; cons - you don't have all the data.

       Line: pros - you have all the data; you can see the distribution (spread, range, clusters, gaps); cons - you don't have the quartiles.

    d. Means are 72.1 and 80.8; medians are 70.5 and 82. Modes are not useful here.

    e. Ranges are 47 vs. 36.

    f. Standard deviations are 11.9 vs. 9.3.

    g. 71% vs. 70%

11. 48 students.

13. a. A situation with outliers.

    b. A situation where you would want to know standard deviation; mean and standard deviation go together well; grades.

15. $\frac{1}{4} \times \frac{2}{8} = \frac{1}{16}$ is the probability of drawing two white circles.

17. Most likely sum is 9, which can be arrived at with 1&8, 2&7, 3&6, 4&5, 5&4, 6&3, 7&2, or 8&1. There are 64 ways to roll the dice, so the probability of rolling a sum of 9 is 8/64 = 1/8.

19. The probability of randomly choosing a number with a zero is 9/90 = 1/10. The probability of randomly choosing a number with a 5 is 18/90 = 1/5.

21. Not fair. Probability of a match is 1/3. You could make it fair by giving player A 2 points if the colors match and player B 1 point if they don't match.

23. It doesn't matter what the first card is, there are 12 out of 51 ways to get a matching suit with the second card, thus 12/51 = 4/17.

25. 0.2, or 20%

27. a. $_6C_2 = \frac{6!}{4!2!} = 15$ ways to choose a schedule

    b. 3

    c. 5

CHAPTER 8 Geometry as Shape

SECTION 8.1 Basic Ideas and Building Blocks

1. **a.** A tetromino is the mathematical name for a Tetris piece.

   b. *Faulty definition:* Four squares that touch each other.

   Fixed definition: Four squares on a sheet of graph paper where each square intersects at least one of the others at a whole edge.

3. **a.** The lines are the same length.

   b. The circles are the same size.

   c. Yes.

   d. The segment on the far right.

   e. Answers will vary. Most likely, a triangle with circles at its vertices will be seen.

   f. It is a 2-dimensional drawing of an impossible 3-dimensional figure.

5. 12 different rays.

7. **a.** False. The lines could be skewed.

   b. True. Given any two parallel lines, there is a plane that will contain both lines.

   c. True. The lines contain three distinct sets of points: those exclusively on one line, those exclusively on the other line, and the point of intersection. These points determine the plane in which the lines lie.

   d. False. It could be skewed.

   e. False. Think of two lines on the plane determined by this sheet of paper and a third line perpendicular to this plane.

   f. False. Two planes cannot intersect at a point.

9. Several answers are possible. Examples are given.

   a. $\angle ABG$ and $\angle GBE$; $\angle AEB$ and $\angle EAB$

   b. $\angle FEG$ and $\angle GEB$; $\angle AGB$ and $\angle BGE$

   c. $\angle ABC$ and $\angle EBD$

   d. $\angle ABC$ and $\angle ABG$

11. **a.** 24 times   **b.** 9:00 and 3:00   **c.** 150° and (210°)

    d. 75°   **e.** 27°   **f.** Answers will vary.

13. Answers will vary.

15. a.   **b.**   **c.**

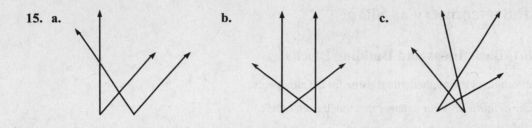

17. Answers may vary. A possible letter "e" for each font is given below.

19. These students are seeing that the rays are longer, but not looking at the opening between the rays.

SECTION 8.2 Two-Dimensional Figures

1. Answers will vary.

3. **a.** The triangle and the kite are most alike, but each pair has similarities.

   b. There are several properties common to all three figures (these are not underlined). You could make and argument for similarities between the second and third figures or for similarities between the first and third figures.

   c. The first and third figures are most similar: regular polygons; multiple pairs of parallel sides; all angles are obtuse.

5. **a.** 20 triangles

   b. 11 rectangles

   c. There are many, for example, trapezoids of different size, parallelograms of different size, pentagons (convex and concave), hexagons (convex and concave), and polygons with more sides.

   d. Answers will vary.

   e. Answers will vary.

7. **a.**

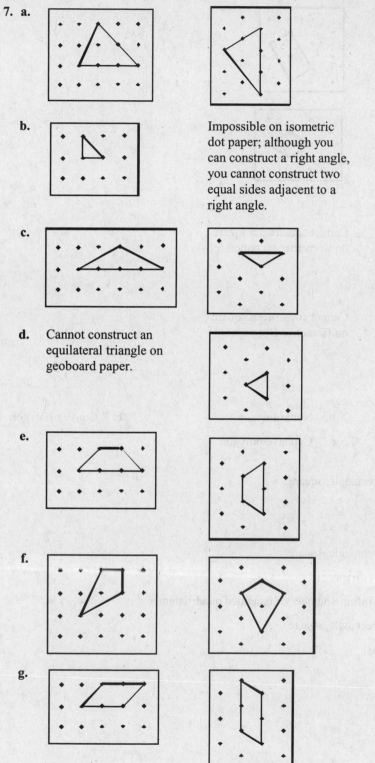

b. Impossible on isometric dot paper; although you can construct a right angle, you cannot construct two equal sides adjacent to a right angle.

c.

d. Cannot construct an equilateral triangle on geoboard paper.

e.

f.

g.

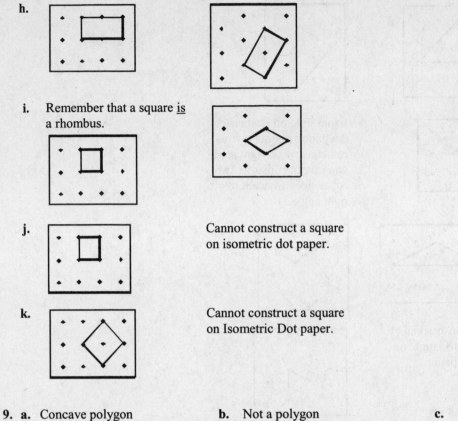

h.

i. Remember that a square _is_ a rhombus.

j. Cannot construct a square on isometric dot paper.

k. Cannot construct a square on Isometric Dot paper.

9. a. Concave polygon   **b.** Not a polygon   **c.** Convex polygon

d. Not a polygon   **e.** Concave polygon

11. a. Parallelogram, rhombus, rectangle, square.

b. Rectangle, square.

c. Square, rhombus.

d. Parallelogram, rhombus, rectangle, square.

e. Rectangle, square, isosceles trapezoid.

f. Rectangle, square, and an infinite number of unnamed quadrilaterals

g. Parallelogram, rhombus, rectangle, square.

h. Kite, unless it is a rhombus.

i. Rhombus.

j. Rectangle.

13. a.

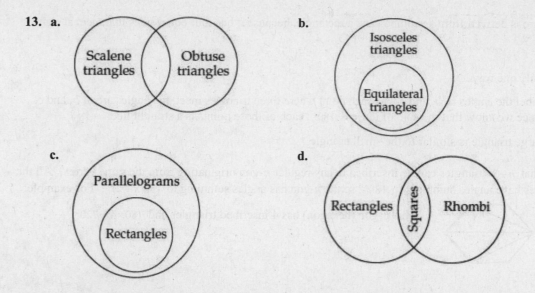

b.

c.

d.

15. A letter "U" (assuming you connect H with A).

17. a. $\left(\dfrac{0+4}{2}, \dfrac{4+10}{2}\right) = (2,7)$ b. $\left(\dfrac{3+7}{2}, \dfrac{4+12}{2}\right) = (5,8)$

19. a. (0, 0), (6, 0), (6, 6), and (0, 6)

b. The other two vertices are on the horizontal line going through $(7,-2)$ and are equidistant from $(7,-2)$; for example, $(5,-2)$ and $(9,-2)$.

c. The coordinates are (10, 0) and (0, 10) or (10, 0) and (0, −10).

d. Answers will vary.

21. If the definition does not include both parts, then there could be more than one kind of regular n-gon for a given length of sides. For example, a regular four-sided figure could be a square or any variety of rhombus. A hexagon could be convex or concave.

23. Answers will vary.

25. a. You can make many different hexagons, both concave and convex, that have all sides equal but not all angles equal. Some have symmetry, some don't.

b. Many possibilities.

c. Many possibilities, all are concave.

d. Many shapes are possible. All must have four consecutive right angles.

e. Two possibilities: both have five 90° angles and one 270° angle. One kind has reflection symmetry and the other doesn't.

f. Yes, squares and rectangles are trapezoids.

g. None.

h. No.

i. Yes, many possibilities.

27. A square is derived from a rhombus and a rectangle, because it has four equal sides that meet at right angles.

29. **b.** Only one way.

   c. Label the angles a, b, and c. At each point where three triangles meet, the angles are a, b, and c. Since we know that the sum of these is $180°$, each of those points is a straight line.

   b. Large triangle is similar to the small triangle.

31. Note that $n-2$ triangles can be inscribed in any regular n-gon originating from the same vertex. All the triangles have angles summing to $180°$, so the n-gon has angles summing to $180(n-2)°$. For example:

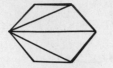 This 6-gon (hexagon) has 4 inscribed triangles and $180 \times 4 = 720°$.

33. ***One way:*** Use a straight edge and compass. Draw any line through the circle, call the points where it crosses the circle A and B. Using the compass, find the perpendicular bisector of AB, call the points where this meets the circle C and D (notice that CD is a diameter of the circle). Now bisect CD and the midpoint of CD will be the center of the circle.

35. **a.** A 3RIT is a figure made from 3 right isosceles triangles joined together so that when a side meets a side, there is no overlap.

   b. Here are some examples:

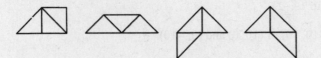

   c. They are congruent. You can flip and rotate one to have it in the same orientation as the other.

   d. Here are some examples:

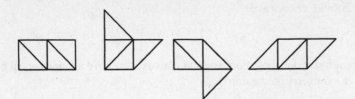

37. $A(0,0)$ $F(8,4)$
   $B(8,0)$ $G(8,8)$
   $C(4,4)$ $H(4,8)$
   $D(6,2)$ $I(2,6)$
   $E(6,6)$ $J(0,8)$

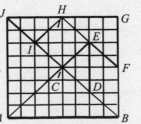

39. a. Orient the triangle so that its base is on the x-axis with one vertex at the origin. Let the other two vertices be called $(4a,0)$ and $(2a,2b)$. The midpoints of the isosceles sides are (a,b) and $(3a,b)$. The base measures $d = \sqrt{(4a-0)^2 + (0-0)^2} = \sqrt{16a^2} = 4a$ and the distance of the segment connecting the midpoints is $d = \sqrt{(3a-a)^2 + (b-b)^2} = \sqrt{4a^2} = 2a$. Therefore the segment is one-half the length of the base.

    b. True for all triangles, you can prove it the same way as in part (a).

41. a. There are many ways to show this. One way is to make use of the Pythagorean Theorem.

    b. One diagonal has slope 1 and the other has slope -1. $(1)(-1) = -1$, so the lines are perpendicular. (Note that this assumes that the square is oriented with its sides running vertically and horizontally.)

43. Answers will vary, but might include: It has four sides, it has more than one right angle, all the sides are congruent, opposite sides are parallel.

45. The correct answer is d.

47. a. They don't have the same number of sides and angles.

    b. All the sides and angles are congruent.

SECTION 8.3 Three-Dimensional Figures

1. a. a cube, a rectangular prism are two possibilities

    b. pentagonal pyramid

3. Answers will vary. There are several correct answers.

    a. $\triangle ABC$ **b.** Point A **c.** $\overline{AB}$

5. a. Octagonal prism   **b.** Triangular prism

    c. Rectangular prism   **d.** Right cylinder

7. a. They have at least one pair of parallel sides.

    b. Prisms have two parallel bases that are congruent polygons.

    c. At least one simple closed curve base.

9. Figures on isometric dot paper should look the same as figures in book.

11. a. rectangle   **b.** rectangle   **c.** rectangle   **d.** rectangle   **e.** rectangle

13. $2n$ vertices; $n+2$ faces; $3n$ edges.

15. Answers will vary.

17. Yes. If all the edges of the base are of different lengths, then the triangular faces will not be congruent.

19.

Base of prism	Number of diagonals
Triangular	$3 \cdot 2 = 6$
Square	$4 \cdot 4 = 16$
Petagonal	$5 \cdot 6 = 30$
n-gon	$\dfrac{v(v-1)}{2}$

21. The sides will always be isosceles triangles.

23. There are nine.

25. The net in a.

27. Answers will vary.

29. Figures P and Q both have 6 faces. The correct answer is a.

31. The correct answer is d.

33. He needs one circle for the base of each cone and two circles for the bases of each cylinder. If he makes 4 cones, he has $10 - 4 = 6$ circles left to make cylinders. So, he can make $6 \div 3 = 2$ cylinders. The correct answer is c.

CHAPTER 8 REVIEW EXERCISES

1. Answers will vary.

3. **a.** False. Consider two lines that lie on this paper and one that is perpendicular to the paper.

   b. False. They could be skew.

5. Answers will vary. One option is to say that a triangle is a shape made by three line segments such that the endpoints of each segment touch the endpoints of another segment.

7. Answers will vary. One option is to draw a quadrilateral and one diagonal, thus making two triangles. Knowing that the sum of the angles of a triangle is $180°$ leads to the conclusion that the sum of the angles of the quadrilateral is $360°$.

9. a.   **b.**   **c.**   **d.**   **e.**

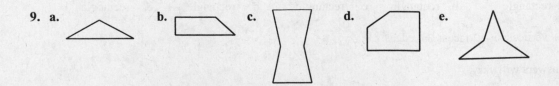

11.

The first figure	The second figure
Hexagon	Hexagon
Concave	Convex
Two sides parallel	Three pairs of parallel sides
Two pairs of congruent sides	
4 acute angles	0 acute angles
2 reflex angles	0 reflex angles
0 right angles	2 right angles
0 obtuse angles	4 obtuse angles
3 pairs of congruent angles	3 pairs of congruent angles
1 line of symmetry	2 lines of symmetry
no rotation symmetry	180° rotation symmetry

13. a. (1) Kiwis are composed of a pentagon with a line segment protruding from one vertex.

    (2) The protruding segment is perpendicular to the side the segment would intersect if it were extended.

    (3) Kiwis have 2 adjacent right angles, which is equivalent to saying that each Kiwi has 2 parallel sides that meet a third side at 90° angles. The first not-Kiwi has 6 sides. The second not-Kiwi has the protruding segment not perpendicular to the opposite side. The third not-Kiwi has the line segment inside the pentagon; the fourth not-Kiwi does not have 2 adjacent right angles.

  b. Figure (a) is not a Kiwi - the protruding segment would not intersect the opposite side if extended. Figure (b) is a Kiwi. Figure (c) is not a Kiwi - the line segment is partially inside and partially outside the pentagon. Figure (d) is not a Kiwi - t does not have 2 adjacent right angles.

15. No, because it implies that there are some rectangles that are not parallelograms.

17. (2.5, 7.5)

19. (6, 6) and (12, 6), (2, –6) and (4, 6), (6, 0) and (12, 0), (–2, 0) and (4, 0).

21. They have a base and an apex.

23.

25.

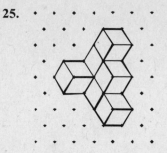

27. Answers will vary. One response: a line segment connecting two nonadjacent vertices.

29.

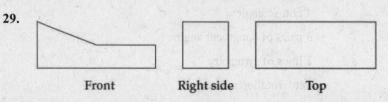

Front   Right side   Top

CHAPTER 9 Geometry as Measurement

SECTION 9.1 Systems of Measurement

1. Answers may vary. Following are examples for the given objects.

    a. surface area, volume, amount of pollutants, temperature of the water (at various levels), depth

    c. height, surface area of sides (for painting), surface area of roof (for shingles), surface area of windows, surface area of floors, ratio of area of windows to area of floors, ratio of area of windows to area of floors (to determine adequacy of ventilation)

3. **a.** 0.5   **b.** 45   **c.** 0.670   **d.** 3600   **e.** 0.450   **f.** 35,000   **g.** 0.024

5. *Mental math:* Answers will vary.

    One possible response: Each yard is a little less than $3\frac{1}{2}$ inches less than a meter, so 400 yards would be about $3\frac{1}{2}$ inches $\times 400 = 1400$ inches less than 400 meters. Then notice that 1400 inches ≈ 1440 inches. Since 1440 inches $= 120$ feet $= 40$ yards, we would expect the answer to be a few yards less than 40 yards.

    Actual calculation: ≈ 37 yards.

7. The first estimate is "off" by 5 feet out of 105 feet, which is a little less than 5%. The second estimate is "off" by 5 feet out of 15 feet, which is about a 33% difference.

9. 1 liter is the volume in a cube that is 10 cm by 10 cm by 10 cm (or 1000 cm^3). A kilogram is the mass of 1 liter of water at 4 degrees Celsius.

11. 30 days

13. Answers will vary.

15. **a.** kilometers

     b. gram; kilogram

     c. liter

17. **a-b.** Answers will vary.

19. Answers will vary.

21. The answer cannot be determined unless you know the length of the train.

23. Same amount.

25-29. Answers will vary.

31. Answers will vary.

33. Jeff has 192 fluid ounces of juice. He can fill 192 fl. ounces $\div$ 8 fl. ounces $= 24$ glasses.

35. 368 centimeters = 300 cemtimeters + 68 centimeters = 3 meters + 68 centimeters. The correct answer is a.

37. 60 ceintimeters = 600 millimeters; 600 millimeters ÷ 6 millimeters = 100. The correct answer is c.

SECTION 9.2 Perimeter and Area

1. a. $P = 69$ mm, $A = 195$ mm^2 **b.** $P = 80$ mm, $A = 148.5$ mm^2

   c. $P = 6$ cm, $A = \sqrt{3} \approx 1.7$ cm^2 **d.** $P \approx 37.68$ cm, $A \approx 113.04$ cm^2

   e. $P = 40$ cm, $A = 72$ cm^2 **f.** $P = 35.8$ cm, $A \approx 51.9$ cm^2

3. a. 15 square units   **b.** $19\frac{1}{2}$ square units   **c.** 29 square units

5. Approximately 11.2 feet. Arc length = $\frac{128}{360}\pi(10)$.

7. Strategies will vary.
   a. 175 square centimeters
   b. 180 square centimeters

9. 144 square inches

11. About 8-9 inches.

13. Approximately 45.84 square meters.

15. 44.4% = 4/9 of the square.

17. a. Answers will vary. The curved line traces the top half of one circle and the bottom half of another circle. These circles have their centers along a diameter of the larger circle and radii that are half the radius of the larger circle.

   b. πr

19. 312 bricks.

21. The diameter of the tree is approximately 16.4 feet.

23. Answers will vary.

25. a. Answers will vary.

   b. The area of each piece is 2 square units.

27. Answers will vary according to the person's height. If you assume 1 quarter is 2.5 cm long, the value of a line of quarters 170 cm long is $17. If you assume 1 nickel is 2 mm tall, then the value of a stack of nickels 170 cm high is $42.50.

29. 78 triangles.

31. No, it will not fit.

33. 24,450 miles. 489 miles is $\frac{7.2}{360}$ of the earth's circumference.

35. Answers will vary.

37. Answers will vary.

39. This is counterintuitive: Let d be the diameter of the ball. The height of the can is $3d$, but the distance around the can is 3.14d!

41. Ratio of corresponding sides is 2:1. Ratio of areas is 4:1.

43. **a.** 10π cm

    b. Not possible, because we don't know the height of the part that is cut out.

    c. $34 + 3\pi$ inches.

45. **a.** 25 square inches   **b.** Not possible

    c. Answers will vary. If the rectangle is 4 centimeters × 5 centimeters, $P = 18$ centimeters. If the rectangle is 10 centimeters × 2 centimeters, $P = 24$ centimeters.

    d. 180 feet by 90 feet.

47. The rectangle is getting bigger in its length and width. The area of the new rectangle is $1.5^2 = 2.25$ times the area of the old rectangle.

49. A population of 300 million would give an area of 581 square miles, which is about 1/3 the size of Rhode Island!

51. The computation yields the almost unbelievable answer of 2.5 square feet per person. If we make a rectangle that is 5 feet long, it would have to be 6 inches wide. This number helps us to understand why as many as 1/3 of the people died during the voyage. Other accounts of slave ships report that each male had a space about 16 inches wide.

53. For the perimeter we are measuring linear distances (the perimeter could be stretched into a line); and for area we are filling in the space with squares.

55. Answers will vary depending on the density of the grass.

57. **a.** If we assume that each hamburger is 1/4 pound and a weight of 100 pounds, that person could each 400 hambers a day.

    b. If we take a cereal box that ha 12 ounces of cereal and divide 100 pounds by 3/4 of a pound, we get 133 boxes of cereal!

57. 8 cm by 4 cm

59. 49 ft^2 – 7 ft^2 = 42 ft^2. The correct answr is c.

61. The area of the triangle is half the area of the square. The correct answer is b.

SECTION 9.3 Surface Area and Volume

1. a. S.A. = 1474.3 square inches.
V = 2160 cubic inches.

b. S.A. 35,630 square feet.
V = 312,000 cubic feet.

c. S.A. 242 square feet.
V = 267 cubic feet.

3. a. S.A. = 144π or 452.16 in^2 ; V = 288π or 904.32 in^3.

b. S.A. = 48π or 150.72 sq. in. ; V = 47π or 150.72 in^3.

c. $h = 16 / \pi$ or 5.1m

5. a. S.A. = 47 square feet.
V = 16 cubic feet.

b. S.A. = 695.2 square feet
V = 1221 cubic feet

c. S.A. 1980 square feet.
V = 2985 cubic feet.

7. a. S.A. = 216 square inches ; V = 144 cubic inches

b. Make it 6 inches by 6 inches by 4 inches, S.A. = 168 square inches.

c. Make it a cube. Length of each side would be 5.24 inches, S.A. = 164.75 square inches.

9. If you roll it so the height is 11 inches, the volume is 62.9 cubic inches. If you roll it so the height is 8.5 inches, thevolume is 81.7 cubic inches.

11. a. 60 feet

b. 44.7 feet

c. 34.6 feet

13. a. Predictions and explanations will vary.
The new tent is about 291 cubic feet larger than the old tent.

b. Predictions and explanations will vary.
The new tent uses almost 258 square feet of additional material.

c. Answers will vary.

15. About 105,000 sheets. The shelves can hold about 42% of the yearly purchase.

17. $\approx$ 34 feet.

19. **a.** Stack them in two layers with each layer a square, 2×2 .

   b. Stack them end to end in one layer with space between each block.

21. Answers will vary.

23. Because 1 cubic meter is a cube that is 1 meter by 1 meter by 1 meter, or 100 centimeters by 100 centimeters by 100 centimeters, we find the answer by: $100 \times 100 \times 100 = 1,000,000$ cubic meters.

25. **a.** S.A. of small prism : S.A. of large prism :: 52 : 468 :: 1 : 9. The surface area is 3 times larger in 2 dimensions.

   b. Volume of small prism : volume of large prism :: 24 : 648 :: 1 : 27. The volume is 3 times larger in 3 dimensions.

27. 20 centimeters $\times$ 20 centimeters $\times$ 10 centimeters.

29. **a.** Answers will vary.

   b. The $167' \times 60'$ building would require about 8% more bricks than the square building. The U-shaped building would require 82% more bricks.

   c.   8 gallons, 9.1 gallons, 14.6 gallons.

31. **and** 33. Answers will vary.

35. Answers will vary. Depends on measured dimensions of a ping-pong ball and how you intend to pack the balls.

37. There are other considerations. A cylindrical shape might require less packing material, but there will be wasted space when many of the product is packed into boxes for shipping to stores.

39. $3^2\pi \times 10 = 90\pi \approx 283$ cubic inches

CHAPTER 9 REVIEW EXERCISES

1. Answers will vary.

3. **a.** 34,000 cm   **b.** 345 g   **c.** 2750 ml

5. Attributes include surface area, perimeter, temperature, depth, and average depth.

7. Area = 6 square inches.

9. So the dimension is $31.83\,\text{cm} \times 31.83\,\text{cm}$.

11. 200 cm on a side, which is equivalent to a square that is 2 meters on a side.

13. **a.** There is empty space between the pennies, because circles do not tessellate.

    b. 25,600 pennies

15. Answers will vary. Two responses: (1) Because 12 inches = 1 foot, 144 square inches = 1 square foot. (2) One could also draw a square that measures 1 foot by 1 foot; thus the area is 1 square foot. When we convert to inches, the same square is 12 inches × 12 inches = 144 square inches.

17. $37\frac{1}{2}\%$

19. One ream of paper measures 8.5 inches by 11 inches by 2 inches. Thus all the forms will take up about 433,000 cubic feet. A cube that measures 76 feet on a side would be able to contain all the paper. In more conventional terms, a building 233 feet by 233 feet by 8 feet would be needed to contain all the paper.

21. **a.** 88.9 cubic yards

    b. 97.8 cubic yards

23. 25,600 cubes.

25. Surface area of the prism is $248\,\text{cm}^2$.
    Surface area of the cylinder is $225\,\text{cm}^2$.
    Volume of the cylinder is $246\,\text{cm}^3$.
    Volume of the prism is $240\,\text{cm}^3$.

CHAPTER 10 Geometry as Transforming Shapes

SECTION 10.1 Congruence Transformations

1. Each point of the figure has been moved 2 units to the right and 3 units down. Also, the figure has been moved approximately 3.6 units along a vector that makes a 55° angle with the *x*-axis.

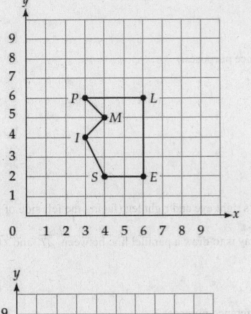

3.

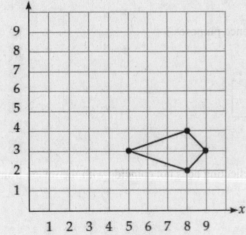

5.

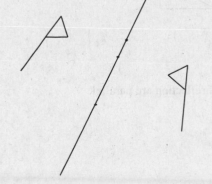

Each point is the same distance from the line as its reflected image.

7.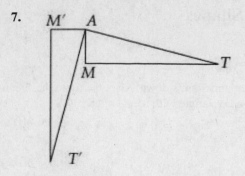

(The original figure is shown for reference purposes.)

9.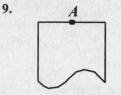

11. Place the mirror so that it bisects Molly's right eye and right leg (facing the left side of Molly).

13. There are many ways to do this. One way is to draw a parallel line between $\overline{AT}$ and $\overline{OG}$ and show that the figures are reflections over this line.

15. a. Answers will vary.

    b. (The original figure is shown for reference purposes.)

    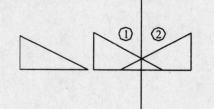

    c. The image from translating and then reflecting does not coincide with the image from reflecting and then translating.

      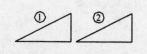

17. They will commute when the translation and the line of reflection are parallel.

19. a.

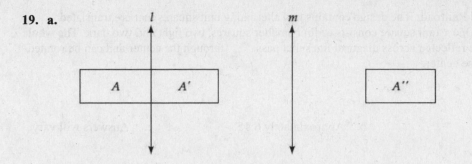

(The original figure is shown for reference purposes.)

c. The images are not the same. The image is located to the left of the original figure by a distance that is approximately equal to the distance between lines *l* and *m*.

21. a.   (The original figure is shown for reference purposes.)

b. Reflect the figure across line *l*, then rotate the image 90° counterclockwise about point *D*.

23. a. Reflect $\overline{PR}$ across $\overline{RM}$.

b. Reflect △*PMR* across $\overline{RM}$.

25. a.   **b.**

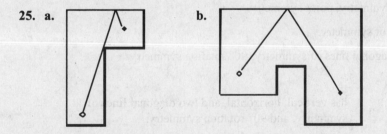

27. Answers will vary. Some possibilities are given.

a. Baby Blocks: If we see the figure as composed of rhombuses, each rhombus can be mapped onto a neighboring rhombus by a 60° rotation.

b. Broken Windows: The top row can be translated onto the second row. A similar translation can be described with respect to columns.

c. Cross Roads: There are horizontal, vertical, and diagonal lines of reflection through the center of the design, and it can be rotated 90°, 180°, or 270° onto itself.

d. Pinwheel: The design can be rotated 90°, 180°, or 270° onto itself. Each white triangle can be rotated 90° and then translated onto another white triangle.

    e. Underground Railroad: The design contains two alternating unit squares that are translated diagonally. One   unit square consists of four smaller squares, two light and two dark. The whole design can be reflected across diagonal lines that pass   through the center and can be rotated 180° about the center.

29. Answers will vary.

31. a. 9 o'clock.   **b.** Approximately 6:15.   **c.** Answers will vary.

33. Answers will vary.

SECTION 10.2 Symmetry and Tessellations

  1. a. $60°, 120°, 180°, 240°,$ and $300°$ rotation symmetry.

     b. Five lines of symmetry through the vertices; $72°$ rotation symmetry.

     c. Horizontal and vertical line symmetry; point symmetry.

     d. Translation symmetry if seen as a pattern extending to the right and left.

  3. a. Vertical line symmetry.

     b. Horizontal, vertical, and diagonal line symmetry; $90°$ rotation symmetry.

     c. No symmetry because of the line segments.

  5. a. Horizontal, vertical, and diagonal line symmetry; $90°$ rotation symmetry.

     b. $90°$ rotation symmetry.

     c. Horizontal and vertical line symmetry; point symmetry.

     d. Horizontal and vertical lines of symmetry.

     e. Horizontal, vertical, and 2 diagonal lines of symmetry; $90°$ rotation symmetry.

  7.

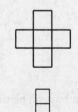

    has vertical, horizontal, and two diagonal lines of symmetry, and $90°$ rotation symmetry.

    has vertical and horizontal line symmetry, and point symmetry.

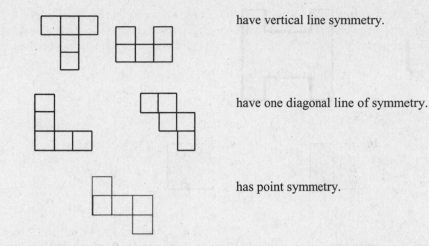

have vertical line symmetry.

have one diagonal line of symmetry.

has point symmetry.

The remaining pentominoes have no symmetry.

9. Notation after semicolon is from p. 590.

   a. rotation; l2

   b. translation, rotation; l2

   c. translation, glide; lg

   d. translation, vertical reflection, horizontal reflection, rotation; mm

   e. translation; 11

   f. translation, vertical reflection , glide; mg

11. Each arrow shape has three vertex points. Verify that the sum of the angles at each vertex point is indeed 360° .

13. All 12 pentominoes tessellate.

15. a. If the unit is one of the C-shapes, then the means of tesselation is a diagonal translation to make one strip of C's, then rotate that entire strip 180 degrees and translate. Then repeat these two steps. If the unit is two C's nested   into each other, then the means is to translate those two C's diagonally to make a long strip. Then repeat.

   b. The unit will be six interlocking hexagons (which form a hexagonal kind of shape). Translate this unit diagonally. Then translate this whole pattern in the other diagonal direction.

   c. The unit is an individual brick. Translate the brick diagonally. Then translate this whole pattern in the other diagonal direction.

   d. This figure is composed of three shapes: a star, a hexagon, and an octagon. One unit consists of the star, two hexagons, and three octagons. The two hexagons are beneath the star and have one common vertex each. The three octagons surround the hexagon that lies to the lower left of the star.

17. It holds.

19. a-g. Answers will vary.

   h. Impossible

21. a. 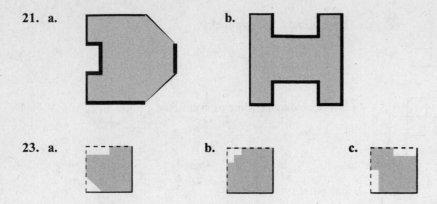 **b.**

23. a.   **b.**   **c.**

25. Answers will vary.

27. All three have the same symmetry: translation, vertical reflection, horizontal reflection, rotation; mm

29. a. 90-degree rotation symmetry and 4 lines of reflection symmetry

   b. 4 lines of symmetry and 90-degree rotation symmetry.

   c. Answers will vary.

31. and 33. Answers will vary.

35. Answers will vary. Possible responses: MATH, WAIT.

37. Answers will vary.

39. All the shapes tessellate.

41. a. False. Justifications may vary.   **b.** False. Justifications may vary.

SECTION 10.3 Similarity

1. a. 13.5 cm   **b.** 2.2 cm

3. Yes, the angles are the same and the sides of the figure to the right are twice as long as the sides of the smaller triangle.

5. a. Corresponding sides of 8×11 and 4×6 rectangles are not proportional.

   b. Answers will vary.

7. 1 inch = 275 miles.

9. Find the height of an object, the length of its shadow, and the length of the shadow of the building. Then write and solve the proportion:

$$\frac{\text{Height of object}}{\text{Length of shadow}} = \frac{\text{Unknown height of building, } x}{\text{Length of building's shadow}}$$

11. Answers will vary.

13. a. The smaller congruent shapes look like the scales on a reptile's body.

   b. Answers will vary.

   c. Answers will vary.

   d. Must be either a parallelogram or have one side perpendicular to the bases. Also, it needs two congruent sides.

   e. There are at least two rectangular tilings and one tiling with a composite unit that looks like a "+" sign.

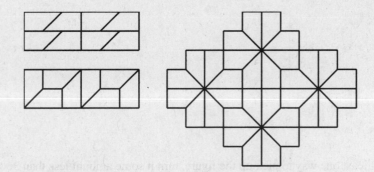

15. It appears that triangles B, C, and E are all similar.

CHAPTER 10 REVIEW EXERCISES

1.

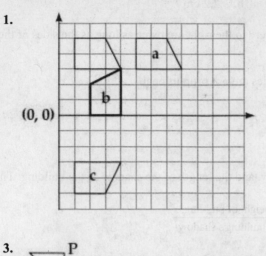

3.

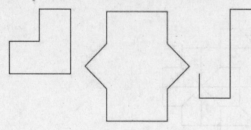

5. The only one-step solution is to rotate the figure 90° counterclockwise about the point that is 3 units to the right and 3 units below the bottom right vertex of the top trapezoid. There are several two-step solutions; for example, translate 6 units down and then rotate 90° counterclockwise.

7. By including glide reflection, we have the theorem that any figure can be moved to any other spot on a plane in *exactly* one move.

9. There are many possibilities. One valid figure is given for each case.

11. a. It means that there is at least one way to pick up the figure, turn it some amount less than 360°, and lay it down so it fits back exactly on itself.

b. It means that there is at least one line that you can draw through the figure such that if you fold the figure along that line, the half of the figure on one side of the line will fit exactly on the other half.

13. a.   There is more than one possibility. For example, the figure at the left can be seen as the unit. In this case, the means by which it is repeated is a translation to the right. On the other hand, the figure on the right can also be seen as the unit. In this case, the means by which it is repeated is also a translation to the right.

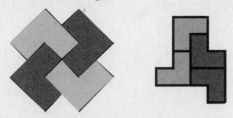

b.   In this case, the unit consists of the five chevrons shown below. The means by which it is repeated is translation in a diagonal direction.

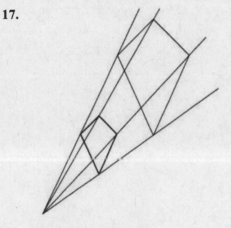

15. $\dfrac{4}{8} = \dfrac{x}{56} \implies 8x = 224 \implies x = 28$ cm

17.

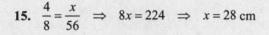